ture notes in pure and applied mathematics

classification theory of semi-simple algebraic groups

I. Satake

38

Classification Theory of Semi-Simple Algebraic Groups

Lecture Notes in Pure and Applied Mathematics

1. *N. Jacobson,* Exceptional Lie Algebras
2. *L.-Å. Lindahl* and *F. Poulsen,* Thin Sets in Harmonic Analysis
3. *I. Satake,* Classification Theory of Semi-Simple Algebraic Groups

Other volumes in preparation

Classification Theory of Semi-Simple Algebraic Groups

I. SATAKE
University of California, Berkeley

With an Appendix by

M. SUGIURA
University of Tokyo

Notes Prepared by Doris Schattschneider
Moravian College, Bethlehem, Pennsylvania

MARCEL DEKKER, INC. New York 1971

These Lecture Notes in Mathematics are produced directly from the author's typewritten notes. They are intended to make available to a wide audience new developments in mathematical research and teaching that would normally be restricted to the author's classes and associates.

The publishers feel that this series will provide rapid, wide distribution of important material at a low price.

MARCEL DEKKER, INC.
95 Madison Avenue, New York, New York 10016

LIBRARY OF CONGRESS CATALOG CARD NUMBER 70-169183
ISBN NO. 0-8247-1607-8

PRINTED IN THE UNITED STATES OF AMERICA

Foreword (added on April, 1971)

This is essentially a photographical reproduction of my Lecture-Notes issued at the University of Chicago in 1967. Taking this opportunity of revision, I tried to make it more readable, eliminating misprints and adding a few foot-notes, a new bibliography, an index of terms, and a list of notations. I also included an Appendix written by M. Sugiura of the University of Tokyo, which gives a very efficient way of classifying real simple algebraic groups in simplification of Araki's method. I should like to express here my gratitude to Sugiura for this invaluable addition to my Notes. My thanks are also due to S. Kobayashi, who invited me to join to the new program of Marcel Dekker mathematics series, to a number of my friends for their kind suggestions for improvements of the Notes, especially to Mrs. Doris Schattschneider for her constant assistance, and finally to Mrs. Laura Hurbace for her fine job in typing these intricate materials.

I. S.

Preface

These notes are based on my course on "Classification-theory of semi-simple algebraic groups" given at the University of Chicago in the winter quarter of 1967. Though its primary aim was to give a general idea of the classification-theory, I thought it convenient to include an outline of the basic theory of algebraic groups, in view of the fact that no standard textbook is as yet available. In this part, proofs are often very sketchy, or completely omitted, but references are given to indicate where a more complete proof is to be found. Thus it is hoped that the graduate student with a sound background in algebra can easily seize the main idea without going into too much detail.

I gratefully acknowledge my debt to Mrs. Doris Schattschneider who kindly helped me in taking notes, reading proofs, and elaborating them in the form presented here.

I. Satake

Table of Contents

I. PRELIMINARIES ON ALGEBRAIC GROUPS

This chapter is an exposition of definitions and known results. The bibliographical references following theorems, or titles of sections indicate where more details and proofs can be found. In a few instances, a proof has been sketched here.

§1. Affine algebraic sets

1.1 Definitions ([4] Chap. II and III; [1] Chap. I, §1)

Notation: Ω : universal domain (i.e., a sufficiently large algebraically closed field)

Ω^N: N-dimensional affine space over Ω

$x = (x_1, \ldots, x_N) = (x_i)$: a point in Ω^N

$\Omega[X] = \Omega[X_1, \ldots, X_N]$: algebra of polynomials in N variables with coefficients in Ω

$\Omega(X)$: quotient field of $\Omega[X]$.

Definition: A subset $A \subset \Omega^N$ is called an (affine) algebraic set if there exists a subset $\mathfrak{M} \subset \Omega[X]$ such that $A = \{x \in \Omega^N \mid f(x) = 0$ for all $f \in \mathfrak{M}\}$. (A is denoted as $A(\mathfrak{M})$, the algebraic set determined by $\mathfrak{M}$; we also write $A \leftarrow \mathfrak{M}$).

Let $A = A(\mathfrak{M})$ be an algebraic set, and put

$$\mathfrak{a}(A) = \{f \in \Omega[X] \mid f(x) = 0 \quad \text{for all } x \in A\}.$$

Clearly $\mathfrak{a}(A)$ is an ideal in $\Omega[X]$, containing $\mathfrak{M}$. Since $\Omega[X]$ is Noetherian, there exists a finite set of polynomials $f_1, \ldots, f_r$ in $\Omega[X]$ which generate $\mathfrak{a}(A)$; then $A = A(\mathfrak{a}(A)) = A(f_1, \ldots, f_r)$. Thus the correspondence between an algebraic set and its corresponding ideal $\mathfrak{a}(A)$ is one-to-one; we will use the notation $A \longleftrightarrow \mathfrak{a}$ ($\mathfrak{a} = \mathfrak{a}(A)$).

It is easy to see that if A_1 and A_2 are algebraic sets in Ω^N, with

$A_1 \longleftrightarrow \mathfrak{A}_1$, $A_2 \longleftrightarrow \mathfrak{A}_2$, then $A_1 \cup A_2 \longleftrightarrow \mathfrak{A}_1 \cap \mathfrak{A}_2$, and $A_1 \cap A_2 \longleftarrow \mathfrak{A}_1 + \mathfrak{A}_2$ ($\longleftrightarrow$ does not hold for the latter). In general, if $A_\alpha \longleftrightarrow \mathfrak{A}_\alpha$, α running through any set of indices, then $\bigcap A_\alpha \longleftrightarrow \Sigma\, \mathfrak{A}_\alpha$. Also, if A and B are algebraic sets in Ω^N and Ω^M respectively, with $A \longleftrightarrow \mathfrak{A} \subset \Omega[X]$, $B \longleftrightarrow \mathfrak{b} \subset \Omega[Y]$, then $A \times B$ is an algebraic set in Ω^{N+M} determined by $\mathfrak{A}\Omega[Y] + \mathfrak{b}\Omega[X]$.

Definition: An algebraic set is called irreducible if $A = A_1 \cup A_2$ (A_1, A_2 non-empty algebraic sets) implies $A = A_1$ or A_2. (An irreducible algebraic set is sometimes called a "variety.")

It follows from the remarks above that an algebraic set A is irreducible if and only if $\mathfrak{A}(A)$ is a prime ideal. Also, every algebraic set can be decomposed uniquely as a finite union of irreducible algebraic sets:

$$A = \bigcup_{i=1}^{n} A_i, \quad A_i \text{ irreducible (all } i\text{), and } A_i \not\subset A_j \text{ if } i \neq j.$$

Now, let k be a subfield of Ω. If $\mathfrak{M}$ is a subset of $\Omega[X]$, denote $\mathfrak{M}_k = \mathfrak{M} \cap k[X]$.

Definition: An algebraic set A is k-closed if and only if there exists a subset $\mathfrak{M} \subset k[X]$ such that $A \longleftarrow \mathfrak{M}$. (Equivalently, $A = A(\mathfrak{A}(A)_k)$.)

An algebraic set A is defined over k (we write A/k) if and only if $\mathfrak{A}(A)$ has a basis in $k[X]$. (Equivalently, $\mathfrak{A}(A) = \mathfrak{A}(A)_k \otimes_k \Omega$.)

If A is defined over k, we say that k is a field of definition for A, and A is sometimes called a "k-rational" algebraic set. It is clear from the definition that if A is defined over k, then A is k-closed.

Definition: Let σ be an automorphism of Ω, and A an algebraic set. The conjugate of A by σ is the set $A^\sigma = \{x^\sigma = (x_i^\sigma) \mid x \in A\}$.

The automorphism σ acts on $\Omega[X]$ (by transforming the coefficients of polynomials), and it is clear that if $A \longleftrightarrow \mathfrak{a}$, then $A^\sigma \longleftrightarrow \mathfrak{a}^\sigma$. If A is k-closed, then A^σ depends only on the restriction of σ to k.

Notation: For $k \subset K$, subfields of Ω, let $\bar{k}$ = algebraic closure of k; k^i = inseparable closure of k; k^s = separable closure of k; Aut(K/k) (or Gal(K/k)) the group of automorphisms of K leaving k pointwise fixed.

Proposition 1.1.1: For an algebraic set A, the following conditions are equivalent:

1) A is defined over k^i.

2) A is k-closed.

3) $A^\sigma = A$ for all $\sigma \in \mathrm{Aut}(\Omega/k)$.

3') A is $\bar{k}$-closed, and $A^\sigma = A$ for all $\sigma \in \mathrm{Gal}(\bar{k}/k)$.

(1) => 2) => 3) is almost trivial; 3) => 1) follows from the Lemma of Weil on field of definition.)

Corollary: If A is defined over k^s and A is k-closed, then A is defined over k.

From this proposition we see immediately that if k is a perfect field (i.e., $k = k^i$), then the terms "k-closed" and "defined over k" for an algebraic set are synonymous. Later we will only be concerned with the case of k a perfect field.

From the proposition (or the definition), it is easy to check that if A_1, A_2 are k-closed algebraic sets in Ω^N, then $A_1 \cup A_2$ and $A_1 \cap A_2$ are also; in general, if $\{A_\alpha\}$ is any collection of k-closed sets, then $\cap A_\alpha$ is k-closed. Thus k-closed algebraic sets satisfy the usual topological conditions of closed sets. The topology on Ω^N having as its closed sets the k-closed algebraic sets is called the "Zariski-k-topology" (or "Zariski topology" when $k = \Omega$).

Unless otherwise specified, in all that follows, by "k-open" (resp., "open") and "k-closed" (resp., "closed") sets in Ω^N, we will always mean with respect to the Zariski-k (resp., Zariski) topology.

It should be noted that the Zariski topology on Ω^N does not satisfy the Hausdorff separation axiom; in fact, if O_1 and O_2 are any non-empty k-open subsets of Ω^N, then $O_1 \cap O_2$ is also a non-empty k-open subset. In addition, any (relatively) open subset of an irreducible set A is necessarily dense in A.

1.2 Rational mappings ([4] Chap. IV; [1] Chap. I, §1)

Definition: Let A be an algebraic set in Ω^N.

A polynomial function (defined over k) on A is the restriction to A of a function defined by a polynomial in $\Omega[X]$ (resp., $k[X]$).

A rational function (defined over k) on A is the restriction to A of a function defined by a rational quotient f/g in $\Omega(X)$ (resp., $k(X)$), with g not vanishing identically on each irreducible component of A. (This last condition is equivalent to: if $A = \cup A_i$ is the decomposition of A into irreducible components, and $A_i \longleftrightarrow \mathfrak{L}_i$, then $g \notin \mathfrak{L}_i$, all i).

Notation: We denote by $\Omega[A]$ (resp., $k[A]$) the ring of polynomial functions on A (defined over k), and by $\Omega(A)$ (resp., $k(A)$) the ring of rational functions on A (defined over k).

The ring $\Omega[A]$ can be canonically identified with $\Omega[X]/\mathfrak{A}(A)$, so that it is an integral domain when A is irreducible, and in that case, $\Omega(A)$ is just the quotient field of $\Omega[A]$.

Definition: Let A be an irreducible algebraic set. The dimension of A is the transcendence degree of the field extension $\Omega(A)/\Omega$. (We write

$\dim A = \dim(\Omega(A)/\Omega)$.)

When A is irreducible, and A/k, one has $\Omega[A] = k[A] \otimes_k \Omega$, so that $\dim A = \dim(k(A)/k)$.

<u>Definition</u>: Let A and B be algebraic sets in Ω^N and Ω^M respectively. A <u>polynomial</u> (resp., <u>rational</u>) <u>map</u> φ from A to B is a mapping given by $\varphi = (\varphi_1,\ldots,\varphi_M)$, $\varphi_i \in \Omega[A]$ (resp., $\varphi_i \in \Omega(A)$), $1 \leq i \leq M$. If φ is a rational map from A to B and each φ_i is represented by $f_i/g_i \in \Omega(A)$, and $x \in A$ satisfies $g_i(x) \neq 0$, $1 \leq i \leq M$, then we say that φ <u>is defined at</u> x, and the value of φ at x is $\varphi(x) = (\varphi_1(x),\ldots,\varphi_M(x)) = \left(\frac{f_1(x)}{g_1(x)},\ldots,\frac{f_M(x)}{g_M(x)}\right) \in B$. We say that φ <u>is defined over</u> k (we write φ/k) if $\varphi_i \in k[A]$ (resp., $k(A)$), $1 \leq i \leq M$.

From the definitions, we see that a rational function on an algebraic set A is a rational map from A to $\Omega^1 = \Omega$. It also follows that any rational map φ from A to B is defined on a non-empty open set in each irreducible component of A. In fact, if we denote by A_φ the subset of points of A at which φ is defined, then $A_\varphi = \bigcup A_{i\varphi}$ (A_i the irreducible components of A), and if φ/k, then $A_{i\varphi}$ is a k-open set in A_i (for all i).

<u>Proposition 1.2.1</u>: A rational map φ from A to B is a polynomial map if and only if $A_\varphi = A$.

<u>Definition</u>: We say a polynomial map φ is a <u>birational isomorphism</u> if φ is bijective and φ^{-1} is also a polynomial map.

<u>Notation</u>: If φ is a rational map from A to B, and M is any subset of A, then denote by $\varphi(M)$ the set-theoretic image of $M_\varphi = A_\varphi \cap M$ by φ in B; and denote by $\overline{\varphi(M)}$ the Zariski-closure of $\varphi(M)$ in B. ($\overline{\varphi(M)}$ is called the algebraic image of M by φ.)

If A is a k-closed algebraic set in Ω^N, denote $A_k = A \cap k^N$. (A_k is called the set of k-rational points of A.)

If φ is a rational map from A to B, with φ/k, A and B k-closed, then clearly $\varphi(A_k) \subset B_k$; in particular, if φ is an isomorphism, φ gives a one-to-one correspondence between A_k and B_k. The following proposition sums up some facts relating $\varphi(A)$ and $\overline{\varphi(A)}$.

Proposition 1.2.2: Let A and B be algebraic sets, A irreducible, and φ a rational map from A to B. Then:

1) $\varphi(A)$ contains a set which is relatively open in $\overline{\varphi(A)}$. In fact, if U is any non-empty (relatively) open subset of A, then $\varphi(U)$ contains a subset which is relatively open in $\overline{\varphi(A)}$.
2) $\overline{\varphi(A)}$ is irreducible.
3) If A is defined over k and φ is defined over k, then $\overline{\varphi(A)}$ is defined over k.

If A and B are irreducible algebraic sets, and φ is a surjective rational map from A to B, then there is a natural injection $\Omega(A) \leftarrow \Omega(B)$ given by $\psi\circ\varphi \leftarrow \psi$ $(\psi \in \Omega(B))$. Under this injection, $\Omega(B)$ can be identified with a subfield of $\Omega(A)$, and we make the following definition.

Definition: The degree of φ, denoted deg φ, is the degree $[\Omega(A):\Omega(B)]$, if this is finite; otherwise the degree of φ is zero. We call φ inseparable (resp., separable) if $\Omega(A)/\Omega(B)$ is a purely inseparable (resp., separable) extension.

§2. Affine algebraic groups ([1] Chap. I; [2] exposé 3; [13] Chap. I)

2.1 Definitions

Definition: G is called an (affine) algebraic group if

1) G is an abstract group;

2) G is an algebraic set in Ω^N;

3) The mapping $G \times G \to G$ is a polynomial map.
$(x,y) \to x^{-1}y$

G is defined over k (write G/k) if G as an algebraic set is defined over k, and the mapping in 3) is defined over k.

If G is an algebraic group, then for any fixed $a \in G$, the left (resp., right) translation

$$L_a: x \to ax, \quad x \in G$$

$$(R_a: x \to xa, \quad x \in G)$$

is an automorphism of G with respect to the structure of an algebraic set. Since left translations are transitive, G is a "homogeneous" algebraic set; in particular, G has no "singular" points. [1] These facts are used in the proofs of some of the properties of algebraic groups.

If G is an algebraic group defined over k, then the identity element of G is k-rational, and it is easily seen that G_k is an abstract group.

If G is an algebraic group and G^o is an irreducible component of G containing the identity element, 1, then it can be shown that G^o is the only irreducible component of G containing 1. Further, we have:

Proposition 2.1.1: Let G be an algebraic group defined over k, G^o the irreducible component of G containing 1. Then G^o is a normal, algebraic subgroup of G, defined over k, and $G = \bigcup g_i G^o$, the coset decomposition of G with respect to G^o, is the decomposition of G into irreducible components.

From this proposition, we see that an algebraic group G is irreducible if and only if it is a connected set in the Zariski topology.

(Note: the words "connected" and "irreducible" are <u>not</u> interchangeable for an arbitrary algebraic set A.) Also from this proposition, we see that the dimension of each of the irreducible components of G is the same as dim G^o; thus we have the following

<u>Definition</u>: The <u>dimension of an algebraic group</u> G, denoted dim G, is equal to dim G^o.

<u>Examples of algebraic groups</u>

Ex. 1. $G = \mathbb{G}_a = \Omega$, the "additive" group of Ω.

$\mathbb{G}_a$ is defined by the zero polynomial, i.e., $\mathbb{G}_a = A(0)$. dim $\mathbb{G}_a = 1$.

Ex. 2. $G = \mathbb{G}_m \simeq \Omega^*$, the "multiplicative" group of Ω.

$\mathbb{G}_m \subset \Omega^2$, and $\mathbb{G}_m = A(XY - 1)$. dim $\mathbb{G}_m = 1$.

Ex. 3. $G = SL(n)$, the "special linear group."

$SL(n) \subset \Omega^{n^2}$, and $SL(n) = A(\det(X_{ij}) - 1)$.

Ex. 4. $G = GL(n)$, the "general linear group."

$GL(n) \subset \Omega^{n^2+1}$, and $GL(n) = A(\det(X_{ij})Y - 1)$.

All of these groups are connected (since their corresponding ideals, being generated by an irreducible polynomial, are prime), and all are defined over the prime field. Note that when k is a topological field, then the group $GL(n)_k = GL(n,k)$ becomes a topological group with respect to the natural topology on k^{n^2+1}. With respect to this natural topology, it can be shown that $GL(n,\mathbb{C})$ is connected, $GL(n,\mathbb{R})$ has two connected components, and $GL(n,\mathbb{Q}_p)$ is totally disconnected. Thus, the Zariski (k-)topology and the natural topology should be carefully distinguished.

Ex. 5. Let G_1, G_2 be algebraic groups in Ω^N and Ω^M respectively. Then $G_1 \times G_2 \subset \Omega^{N+M}$ is also an algebraic group, and is called the direct product of G_1 and G_2. If G_1 and G_2 are both defined over k, then $G_1 \times G_2$ is also defined over k.

2.2 Rational homomorphisms and quotient groups

Definition: Let G be an algebraic group defined over k; $H \subset G$ is a k-closed (resp., defined over k) subgroup if

1) H is an abstract subgroup of G, and
2) H is a k-closed subset (resp. defined over k) of G.

Definition: Let G, G' be algebraic groups. A mapping $\varphi: G \to G'$ is called a (rational) homomorphism defined over k if

1) φ is a homomorphism of abstract groups,
2) φ is a rational map defined over k.

Such a mapping φ is also called a k-homomorphism of G into G'; it should be noted that conditions 1) and 2) imply that φ is in fact a polynomial map (since φ is necessarily defined on all of G).

Definition: A rational homomorphism from an algebraic group G into $\mathbb{G}_m$ is called a character of G.

Proposition 2.2.1: Let G, G' be algebraic groups, with G/k, and φ a k-homomorphism of G into G'.

1) $\varphi^{-1}(1) = \ker \varphi$ is a k-closed normal subgroup of G.
2) $\varphi(G) = \operatorname{Im} \varphi$ is a closed subgroup of G', defined over k.
3) $\dim(\operatorname{Im} \varphi) = \dim G - \dim(\ker \varphi)$.

As a special case of part 3) above, we see that if G and G' are connected and have the same dimension, then φ is surjective if and only if $\ker \varphi$ is finite.

Definition: Let G, G' be connected algebraic groups.

A rational homomorphism of G onto G' having finite kernel is called an <u>isogeny</u>.[1)]

If φ is an isogeny from G onto G' then by definition, deg φ = $[\Omega(G): \Omega(G')]$; if G and φ are both defined over k, then it can be shown that deg $\varphi = [k(G): k(G')]$.

<u>Proposition 2.2.2</u>: Let φ be an isogeny from G onto G'. The number of elements in ker φ equals $[k(G): k(G')]_s$, the degree of the maximal separable extension of k(G') contained in k(G). ([4], Chap. IV, 3)

Thus an isogeny φ is bijective if and only if the extension k(G)/k(G') is purely inseparable (i.e., φ is inseparable).

<u>Definition</u>: Let G, G' be algebraic groups. A mapping $\varphi: G \to G'$ is called a (rational) k-<u>isomorphism</u> if φ is a bijective rational k-homomorphism having as inverse a rational k-homomorphism. If G = G', we call φ a k-<u>automorphism</u> of G; the group of k-automorphisms of G is denoted $\mathrm{Aut}_k(G)$.

It is clear from remarks above that if φ is an isomorphism, then deg φ = 1.

<u>Remark</u>: Not every inseparable isogeny is an isomorphism, as is shown in the following example ("Frobenius endomorphism").

Let G be an algebraic group defined over $k = \mathbb{F}_q$, the finite field having q elements. Let φ be the rational mapping from G into itself defined by $x = (x_i) \to (x_i^q) = (x_i)^{(q)}$, $x \in G$. (φ maps G into itself since for each $x \in G$ and $f \in \mathfrak{A}(G)_k$ one has $0 = f(x) = (f(x))^q = f(x^{(q)})$.) The image of k(G) under the injection induced by φ is equal to $k(G)^q$; in fact, if $\psi \in k(G)$, then $(\psi \circ \varphi)(x) = [\psi(x)]^q$, for all $x \in G$. Clearly φ is defined over k, and is an automorphism of G as an abstract group, but φ

is not necessarily an isomorphism since $[k(G): k(G)^q] > 1$ in general.

Quotient groups ([1]-I, §5; [2], 3-05 and exposé 8; [13]-II)

Let G be an algebraic group, defined over k, and N a closed normal subgroup, also defined over k. The following proposition allows us to speak of "the quotient group" G/N, which is an algebraic group defined over k.

Proposition 2.2.3: There exists an algebraic group $\bar{G}$, defined over k, and a surjective homomorphism π, defined over k, $\pi: G \to \bar{G}$ which satisfies:

i) $\ker \pi = N$,

ii) if φ is a rational homomorphism, defined over k, from G into an algebraic group G', satisfying $N \subset \ker \varphi$, then there exists a unique homomorphism $\bar{\varphi}$, defined over k such that $\varphi = \bar{\varphi} \circ \pi$.

By the universal mapping property (ii) we see that $\bar{G}$ is unique up to k-isomorphism; the group $\bar{G}$ is denoted G/N.

One of the key lemmas used to prove this result is due to Chevalley:

Lemma 2.2.4: Let G be an algebraic group, and N a closed invariant subgroup, both defined over k. Then there exists a rational representation φ (defined over k) of G having N as kernel ([2], 4-04).

If G is a connected algebraic group, then k(G/N) can be identified with a subfield of k(G) (since π is surjective). The group N acts on the field k(G) by $\psi \to \psi \circ R_a$, $a \in N$.

The following proposition characterizes G/N.

Proposition 2.2.5: k(G/N) is the subfield of k(G) consisting of all N-invariant functions; i.e., $k(G/N) = \{\psi \in k(G) \mid \psi \circ R_a = \psi, \text{ all } a \in N\}$.

2.3 Linear algebraic groups

Definition: A (finite dimensional) vector space V is said to be defined over k (or have k-structure) if

1) V is a vector space over Ω,

2) there is a given a subset V_k of V which is a vector space over k, spanned over k by an Ω-basis of V, (hence one has $V = V_k \otimes_k \Omega$).

Throughout, unless otherwise specified V will be defined over k; and we fix the k-structure once and for all. If K is another subfield of Ω, $k \subset K \subset \Omega$, then defining $V_K = V_k \otimes_k K$, we see that V has the natural structure of a vector space defined over K.

Notation: GL(V) is the group of non-singular linear transformations of V, GL(V) is called the general linear group of V.

If we take a basis $(e_1, \ldots, e_n)$ of V_k over k, then with respect to this basis, there is a natural isomorphism ρ: $GL(V) \simeq GL(n)$. If a second basis of V_k is taken and ρ' is the isomorphism of GL(V) into GL(n) with respect to it, then it is easy to see that $\rho = \varphi \circ \rho'$ where φ is an inner automorphism of GL(n), defined over k (given by the change of basis matrix). Thus the identification of GL(V) with GL(n) is unique up to an inner automorphism defined over k.

Definition: A subgroup $G \subset GL(V)$ is a linear algebraic group, defined over k (write G/k), if under the isomorphism ρ described above, $\rho(G)$ is an affine algebraic group defined over k.

Notation: If G is a linear algebraic group, denote by G_k the inverse image of $\rho(G)_k$ by ρ. (Clearly G_k is uniquely determined.) We call the elements of G_k the k-rational points of G.

The following proposition, easily proved, shows that linear algebraic groups and affine algebraic groups are essentially the same.

Proposition 2.3.1: Any affine algebraic group is isomorphic to a linear algebraic group. ([2], 4-03).

Examples

1. $SL(V) = \{g \in GL(V) | \det g = 1\}$ (special linear group of V).

   The sequence of mappings

   $$1 \to SL(V) \to GL(V) \xrightarrow{\det} \mathbb{G}_m \to 1$$

   is exact, and it can be shown that $\mathbb{G}_m$ and det satisfy the universal mapping property (ii) of Prop. 2.2.3. Thus $\mathbb{G}_m = GL(V)/SL(V)$.

2. Let Z = center of GL(V) = scalar multiplications. The quotient group GL(V)/Z is denoted PL(V), called the projective linear group of V. The realization of PL(V) as a matrix group is as follows: let V^* = dual of V, and π the mapping of $GL(V) \to GL(V^* \otimes V)$ defined by $g \xrightarrow{\pi} {}^t g^{-1} \otimes g$, $g \in GL(V)$. Then $\ker \pi = Z$, and the pair $\pi(G)$, π satisfies (ii) of Prop. 2.2.3, so that $\pi(G) = PL(V)$.

3. Let B be a non-degenerate bilinear form on $V \times V$. (Recall that two bilinear forms B, B' are equivalent, $B \sim B'$, if and only if there exists an element $g \in GL(V)$ such that $B'(x,y) = B(gx,gy)$ for all $(x,y) \in V \times V$. If $(e_1,\ldots,e_n)$ is a basis for V_k, and B is identified with the matrix $(B(e_i,e_j))$, then $B \sim B'$ if and only if $B' = {}^t g B g$ for some $g \in GL(n)$.)

   If B = S is a symmetric form, then define $O(V,S) = \{g \in GL(V) | S = {}^t g S g\}$, the orthogonal group of V with respect to S; $SO(V,S) = O(V,S) \cap SL(V)$, the special orthogonal group of V with respect to S.

   If B = A is an alternating form, then define

$Sp(V,A) = \{g \in GL(V) \mid A = {}^t g A g\}$, the <u>symplectic group</u> of V with respect to A.

We say B is defined over k (write B/k) if the matrix $(B(e_i,e_j)) \in GL(n)_k$; clearly this does not depend on the choice of the basis $(e_1,\ldots,e_n)$ of V_k. If B/k, then the three groups just defined are all defined over k.

If two bilinear forms B, B' are both defined over k, we will write $B \underset{k}{\sim} B'$ if $B' = {}^t g B g$ with g k-rational. It is known that if char k $\neq 2$, then there is only one equivalence class over k of alternating forms. Thus if A_o/k is an alternating form, we can choose a basis of V_k such that A_o has matrix $\begin{pmatrix} 0 & -1_m \\ 1_m & 0 \end{pmatrix}$, $n = 2m$. (We write $Sp(V,A_o) = Sp(m)$.) If $A \sim A_o$, then $A = {}^t g\, A_o g$, and det(g) is a polynomial function of A, uniquely determined by A; det(g) is called the "Pfaffian" of A relative to A_o. If $g_1 \in Sp(V,A)$, then $A = {}^t g\, A_o g = {}^t(gg_1)A_o(gg_1)$, so $\det(gg_1) = \det(g)$, hence $\det g_1 = 1$. Thus $Sp(V,A) \subset SL(V)$.

Finally, it is known that SO(V,S) and Sp(V,A) are connected algebraic groups (and are simple groups except for SO(V,S) with dim V = 1,2,4).

2.4 <u>Tori</u> ([1], Chap. II; [2], exposé 4; [5])

<u>Definition</u>: An algebraic group G is called a <u>torus</u> if, for some n, there exists an isomorphism φ of G onto $(\mathbb{G}_m)^n$. G is said to be k-<u>trivial</u> (or split over k) if G and φ are both defined over k.

Throughout this section, we will use T to denote a torus, and X = X(T) the group of characters of T. We use additive notation for X (i.e., by definition, $(\chi_1 + \chi_2)(t) = \chi_1(t)\chi_2(t)$ for $\chi_1, \chi_2 \in X$, $t \in T$) so that X is a $\mathbb{Z}$-module.

The group of diagonal matrices D(n) in GL(n) is a torus ($\simeq (\mathbb{G}_m)^n$)

in Ω^{n+1}, with $D(n) \longleftrightarrow \mathfrak{a} = (X_1 \cdots X_n Y - 1)$, and clearly $D(n)$ splits over the prime field. Thus we identify $D(n) = (\mathbb{G}_m)^n$. Let $T = (\mathbb{G}_m)^n$; there are n "canonical" characters χ_i of T defined by: if $x = (x_1, \ldots, x_n) \in T$, then $\chi_i(x) = x_i$. Under the identification $\Omega[T] = \Omega[X_1, \ldots, X_n, Y]/\mathfrak{a}$, the character χ_i is identified with $X_i \pmod{\mathfrak{a}}$, and the function $(\prod_{i=1}^{n} \chi_i)^{-1}$ is identified with $Y \pmod{\mathfrak{a}}$. From this it can be shown that (1) $\Omega[T] = \Omega[\chi_1^{\pm 1}, \ldots, \chi_n^{\pm 1}]$, (2) $\Omega(T) = \Omega(\chi_1, \ldots, \chi_n)$ is a purely transcendental extension of Ω, (3) X is just the subset of monomials $\chi_1^{m_1} \cdots \chi_n^{m_n}$ in $\Omega[T]$. This last fact shows that as a $\mathbb{Z}$-module, $X \cong \mathbb{Z}^n$.

Now let T be any torus and X its character group. As might be expected, there is "duality" between certain submodules of X and subgroups of T; although we restrict to algebraic subgroups of T, most of the usual duality results hold.

If T_1 is a closed subgroup of T, denote by $T_1^{\perp}$ the submodule of X: $T_1^{\perp} = \{\chi \in X \mid \chi(t) = 1, \text{ all } t \in T_1\}$. Similarly, if X_1 is a submodule of X, denote by $X_1^{\perp}$ the closed subgroup of T defined by:

$$X_1^{\perp} = \{t \in T \mid \chi(t) = 1, \text{ all } \chi \in X_1\}.$$

We have the following

<u>Proposition 2.4.1</u>:

1) $T_1 \to T_1^{\perp}$, $X_1 \to X_1^{\perp}$ define reciprocal bijections between the set of closed subgroups of T and the set of submodules of X satisfying the condition X/X_1 has no p-torsion (p = characteristic of the prime field; if p = 0 there is no condition on torsion).
2) A closed subgroup $T_1 \subset T$ is connected if and only if X/X_1 has no torsion, where $X_1 = T_1^{\perp}$. T_1 is a torus if and only if it is connected.

3) $X(T_1) = X/X_1$, $X(T/T_1) = X_1$ where $X_1 = T_1^{\perp}$.

Two key lemmas used to prove this proposition are:

Lemma 2.4.2 (Chevalley): Let H be a k-closed subgroup of an algebraic group G/k. Then there exists a rational representation ρ/k of G into GL(V) and a vector $v_1 \in V_k$ such that $H = \{g \in G | \rho(g)v_1 = \lambda(g)v_1, \lambda(g) \in \Omega\}$. ($\lambda$ is a character on H.) ([2],4-03)

Lemma 2.4.3: Any rational representation of a torus T is completely reducible.

Let T and T' be tori with character modules X, X' respectively. If φ is a homomorphism of T into T', then the mapping ${}^t\varphi$ of X' into X, defined by ${}^t\varphi(\chi') = \chi' \circ \varphi$, $\chi' \in X'$, is a module homomorphism. Conversely, if ψ is a module homomorphism of X' into X, then there exists a rational homomorphism φ of T into T' such that $\psi = {}^t\varphi$. To see this, we can identify T with $(\mathbb{G}_m)^n$ and T' with $(\mathbb{G}_m)^m$; then $X = \{\chi_1,\ldots,\chi_n\}_{\mathbb{Z}}$ and $X' = \{\chi_1',\ldots,\chi_m'\}_{\mathbb{Z}}$, and so there corresponds to ψ an integral matrix (m_{ij}) where $\psi(\chi_i') = \sum_{j=1}^{n} m_{ij}\chi_j$, $1 \leq i \leq m$. φ is defined as the mapping which sends $x = (x_1,\ldots,x_n) \in T$ to the element in T' whose i^{th} coordinate is $\prod_{j=1}^{n} x_j^{m_{ij}}$. It is clear that φ is a rational homomorphism from T into T' and that $\psi = {}^t\varphi$. It is easily checked that this establishes a one-to-one correspondence between homomorphisms of tori and homomorphisms of their character modules.

Proposition 2.4.4: Let φ be a homomorphism from T into T'.

1) $(\mathrm{Im}\ \varphi)^{\perp} = \mathrm{Ker}\ {}^t\varphi$, so φ is surjective if and only if ${}^t\varphi$ is injective.

2) $\mathrm{Ker}\ \varphi = (\mathrm{Im}\ {}^t\varphi)^{\perp}$, (this implies $\mathrm{Im}\ {}^t\varphi$ is a submodule of index q = (power of p) in $\mathrm{Ker}\ \varphi^{\perp}$), so φ is injective if and only if $[X:{}^t\varphi(X')]$ is a power of p.[2)]

If we apply the proposition in the case where dim T = dim T', we see that all of the following statements are equivalent: (i) φ is an isogeny, (ii) φ is surjective, (iii) ${}^t\varphi$ is injective, (iv) φ has finite kernel, (v) ${}^t\varphi$ has finite cokernel. It is easily shown that if φ is an isogeny, then $\deg \varphi = [X: {}^t\varphi(X')]$, so that φ is an isomorphism if and only if ${}^t\varphi$ is an isomorphism.

<u>Example</u>: Let T be defined over $k = \mathbb{F}_q$, and let φ be the Frobenius endomorphism of T into itself (see <u>Remark</u>, after Prop. 2.2.2). If $\chi \in X$, then by definition, $({}^t\varphi(\chi))(t) = \chi(t^{(q)})$ for $t \in T$. If we denote by $\chi^{(q)}$ the character gotten from χ by taking q^{th} powers of its coefficients, we see that $(\chi(t))^{(q)} = \chi^{(q)}(t^{(q)})$. However, $\chi(t)$ is a scalar, so that $(\chi(t))^{(q)} = (\chi(t))^q = (q\chi)(t)$, using additive notation. Thus ${}^t\varphi(\chi^{(q)})$ $= q\chi$, and if χ is defined over $\mathbb{F}_q$, then ${}^t\varphi(\chi) = q\chi$.

We end this section with some results on the field of definition and splitting field of a torus.

<u>Proposition 2.4.5</u>: If T is a torus defined over k, then T splits over k^s.

Let $\Gamma = \mathrm{Gal}(k^s/k)$, and T a torus defined over k. Since there is an isomorphism φ/k^s of T onto $(\mathbb{G}_m)^n$, ${}^t\varphi$ gives an isomorphism of $X((\mathbb{G}_m)^n)$ onto $X(T) = X$; since the canonical characters χ_i of $(\mathbb{G}_m)^n$ are all defined over the prime field, clearly the characters ${}^t\varphi(\chi_i)$ are defined over k^s. Thus every $\chi \in X$ is defined over k^s, and so X has the structure of a Γ-module (as usual, if $\sigma \in \Gamma$, $\chi \in X$, then χ^σ is the character gotten from χ by operating on its coefficients by σ). A similar argument gives the following result:

<u>Proposition 2.4.6</u>: Let T be defined over k. The following statements are equivalent:

1) T is k-trivial,

2) every $\chi \in X$ is defined over k,

3) Γ operates trivially on X.

It should be remarked that if T is defined over k, then in fact T splits over a <u>finite</u> Galois extension K of k. If Γ is replaced by Gal(K/k) in the remarks and 2.4.6 above, they remain true.

We can refine previous results further:

<u>Proposition 2.4.7</u>:

1) Let T be a torus defined over k. If $T_1 \subset T$ and $T_1 \longleftrightarrow X_1 \subset X$ (Prop. 2.4.1), then T_1 is defined over k if and only if X_1 is a Γ-submodule of X.

2) Let T and T' be tori defined over k, and φ a homomorphism from T into T'. φ is defined over k if and only if ${}^t\varphi$ is a Γ-homomorphism; in particular, φ is a k-isomorphism if and only if ${}^t\varphi$ is a Γ-isomorphism.

This proposition leads to the following main theorem on tori:

<u>Theorem 2.4.8</u>: There is a one-to-one correspondence between the category of tori defined over k and the category of finitely generated torsion-free Γ-modules.

To complete the theorem, it must be shown that to each such module X, there corresponds a unique (up to k-isomorphism) torus T such that X(T) and X are isomorphic as Γ-modules. This leads us to the next topic.

§3. K/k-<u>forms</u>

3.1 K/k-<u>forms</u> <u>and</u> <u>one</u> <u>cocycles</u>

<u>Definition</u>: Let k and K be subfields of Ω, $k \subset K$, and G_1 an algebraic group defined over K. A pair (G,f) is called a <u>K/k-form</u> <u>of</u> G_1 if G is

an algebraic group defined over k, and f is an isomorphism of G onto G_1, f defined over K. A pair (G,f) is called a <u>k-form</u> of G_1 if (G,f) is a K/k-form of G_1 for some (sufficiently large) extension K of k.

We are only interested in the case where k is <u>perfect</u>, and K/k is a <u>finite</u> extension, and make these assumptions for the remainder of this section.

Let $\Gamma = \mathrm{Gal}(\bar{k}/k)$ (note $\bar{k} = k^s$).

If G_1 is an algebraic group defined over K and (G,f) is a K/k-form of G_1, then any element $\sigma \in \Gamma$ acts on the coefficients of the polynomial mapping f; let f^σ denote this new mapping. Since G/k, $G^\sigma = G$, so that f^σ is an isomorphism of G onto $G_1^{\ \sigma}$. Define $\varphi_\sigma = f^\sigma \circ f^{-1}$; φ_σ is then a $\bar{k}$-isomorphism of G_1 onto its conjugate $G_1^{\ \sigma}$.

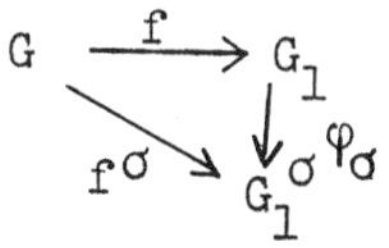

From the definition of φ_σ, it is easy to check that the system of isomorphisms $(\varphi_\sigma)_{\sigma \in \Gamma}$ satisfies the condition

$$(1) \qquad \varphi_\sigma^{\ \tau} \circ \varphi_\tau = \varphi_{\sigma\tau}.$$

Although Γ is an infinite group, the system $(\varphi_\sigma)_{\sigma \in \Gamma}$ is essentially finite, for f is defined over K, and from the definition of φ_σ, it is clear that

(2) φ_σ depends only on the restriction of σ to K, and φ_σ is defined over K' for some finite extension K' of k.

A main result on K/k-forms says that conversely, any such system of mappings $(\varphi_\sigma)_{\sigma \in \Gamma}$ satisfying (1) and (2) necessarily belongs to a K/k-form.

<u>Proposition 3.1.1</u>: Let G_1 be an algebraic group defined over K, and $(\varphi_\sigma)_{\sigma \in \Gamma}$ a system of isomorphisms ($\varphi_\sigma : G_1 \xrightarrow{\cong} G_1^\sigma$), satisfying (1) and (2). Then there exists a K/k-form, (G,f), of G_1 such that $\varphi_\sigma = f^\sigma \circ f^{-1}$ for all $\sigma \in \Gamma$.

We will give a proof of this proposition later; it is true, in fact, if G_1 is any algebraic set. (See A. Weil, "The Field of Definition of a Variety" —A.M.S., Vol. 78.)

<u>Notation</u>: We will denote by (φ_σ) a system of mappings as described in Prop. 3.1.1 (dropping the obvious index set Γ), and call it the system corresponding to the K/k-form (G,f).

<u>Definition</u>: Let (G',f') and (G,f) be k-forms of G_1. We say that (G',f') and (G,f) are <u>isomorphic</u> if there is a k-isomorphism of G onto G'.

Let (G,f) and (G',f') be isomorphic K/k-forms of G_1, (φ_σ) and $(\varphi_\sigma{}')$ their corresponding systems, and ρ a k-isomorphism of G onto G'. Define a mapping ψ by $\psi = f' \circ \rho \circ f^{-1}$; ψ is a K-automorphism of G_1.

$$\begin{array}{ccc} G & \xrightarrow{f} & G_1 \\ \downarrow \rho & & \downarrow \psi \\ G' & \xrightarrow{f'} & G_1 \end{array}$$

If we apply $\sigma \in \Gamma$ to the equation defining ψ, we have (since ρ/k)

$$\psi^\sigma = f'^\sigma \circ \rho \circ f^{-\sigma} = \varphi_\sigma{}' \circ f' \circ \rho \circ f^{-1} \circ \varphi_\sigma{}^{-1} = \varphi_\sigma{}' \circ \psi \circ \varphi_\sigma{}^{-1}.$$

Thus

$$(3) \qquad \varphi_\sigma{}' = \psi^\sigma \circ \varphi_\sigma \circ \psi^{-1}.$$

<u>Definition</u>: Let G_1/K. Two systems of isomorphisms (φ_σ), $(\varphi_\sigma{}')$, where φ_σ and $\varphi_\sigma{}'$ map G_1 onto G_1^σ, and satisfy (1) and (2) are said to be K-<u>equivalent</u> (resp., equivalent) if there exists a K-automorphism (resp., automorphism) ψ of G_1 satisfying (3).

Proposition 3.1.2: Let G_1/K, (G,f) and (G',f') K/k forms (resp., k-forms) of G_1. (G,f) and (G',f') are isomorphic if and only if their corresponding systems (φ_σ) and (φ_σ') are K-equivalent (resp., equivalent).

We have just proved this statement in one direction. Conversely, if (φ_σ) and (φ_σ') are systems which are K-equivalent (or equivalent) under $\psi \in \mathrm{Aut}_K(G_1)$, we know there exist K/k-forms (G,f) and (G',f') of G_1 corresponding to these systems (Prop. 3.1.1). Defining $\rho = f'^{-1} \circ \psi \circ f$, clearly ρ is an isomorphism of G onto G', and $\rho^\sigma = f'^{-\sigma} \circ \psi^\sigma \circ f^\sigma = (f'^{-1} \circ \varphi_\sigma'^{-1}) \circ (\varphi_\sigma' \circ \psi \circ \varphi_\sigma^{-1}) \circ (\varphi_\sigma \circ f) = f'^{-1} \circ \psi \circ f = \rho$, so ρ/k. Thus we have:

Corollary 3.1.3: The k-isomorphism classes of K/k-forms of G_1/K are in one-to-one correspondence with the K-equivalence classes of systems (φ_σ) $(\varphi_\sigma\colon G_1 \xrightarrow{\cong} G_1^\sigma)$ satisfying (1) and (2).

Now assume G_1 is defined over k and K/k is a Galois extension. Then $G_1^\sigma = G_1$ for all $\sigma \in \Gamma$, and the elements φ_σ of a system (φ_σ) corresponding to a K/k-form (G,f) of G_1 are all K-automorphisms of G_1 (since K/k is Galois, $K^\sigma = K$, all $\sigma \in \Gamma$, so f/K implies f^σ/K, hence $f^\sigma \circ f^{-1} = \varphi_\sigma/K$). Condition (1) implies that the system (φ_σ) is a one-cocycle of Γ in $\mathrm{Aut}_K(G_1)$, and condition (3) shows that "cohomologous" and "K-equivalent" are synonymous terms for a pair of systems (φ_σ) and (φ_σ'). Condition (2) implies that the system (φ_σ) can be regarded as a one-cocycle of $\mathrm{Gal}(K/k)$ in $\mathrm{Aut}_K(G_1)$. If we denote by $H^1(\mathrm{Gal}(K/k), \mathrm{Aut}_K(G_1))$ the first cohomology set of $\mathrm{Gal}(K/k)$ in $\mathrm{Aut}_K(G_1)$, then Corollary 3.1.3 becomes:

Corollary 3.1.4: The k-isomorphism classes of K/k-forms of G_1/k are in one-to-one correspondence with the elements of $H^1(\mathrm{Gal}(K/k),\mathrm{Aut}_K(G_1))$.

In the case of a k-form (G,f) of G_1, we see that conditions (1) and

(2) imply that the system (φ_σ) corresponding to (G,f) is a <u>continuous</u> one-cocycle of Γ in $Aut_{\bar{k}}(G_1)$. (By this, we mean that (φ_σ) is a continuous one-cocycle with respect to the Krull topology on Γ, the topology which has as a fundamental system of neighborhoods of the identity the groups $Gal(\bar{k}/K)$, where K/k runs through all finite Galois extensions of k.) If we denote by $H^1(k,Aut(G_1))$ the first cohomology group of Γ in $Aut_{\bar{k}}(G_1)$ (this is the injective limit $\bigcup_K H^1(Gal(K/k),Aut_K(G_1))$, then we have the same correspondence for k-forms.

<u>Corollary 3.1.4'</u>: The k-isomorphism classes of k-forms of G_1/k are in one-to-one correspondence with the elements of $H^1(k,Aut(G_1))$.

3.2 <u>Examples of K/k-forms</u>

Although we have defined K/k-forms for algebraic groups, it is clear that we may define a K/k-form for any algebraic 'object' where "defined over K" and "K-isomorphism" make sense. That is, if M is defined over K, then $(\tilde{M},f)$ is a K/k-form of M if $\tilde{M}$ is defined over k and f is a K-isomorphism of $\tilde{M}$ onto M. To any such K/k form corresponds a system (φ_σ) as defined at the beginning of section 3.1.

In all of the examples which follow, we continue to assume k perfect, K/k finite, $\Gamma = Gal(\bar{k}/k)$.

<u>Example 1</u>: K/k-form of a torus.

Let $G_1 = (\mathbb{G}_m)^n$; G_1 is a torus defined over the prime field. If (G,f) is a K/k form of G_1, then by definition, G is a torus defined over k and split over K. Let $X_1 = X(G_1)$; since elements of the system (φ_σ) corresponding to (G,f) are in $Aut_K(G_1)$, it follows by duality that the elements ${}^t\varphi_\sigma$ are in $Aut_K(X_1) \cong GL(n,\mathbb{Z})$. Since G_1 splits over the prime field, all elements of X_1 are defined over the prime field, and so Γ

acts trivially on X_1. From this (and duality) we have condition (1) implies (1') ${}^t\varphi_\tau \circ {}^t\varphi_\sigma = {}^t\varphi_{\sigma\tau}$. Thus the mapping $\sigma \to ({}^t\varphi_\sigma)^{-1}$ of Γ into $\mathrm{Aut}_K(X_1)$ gives a representation of Γ in $GL(n,\mathbb{Z})$. (Perhaps it should be noted that in this representation, X_1 becomes a left Γ-module; the "usual" action of Γ on the character module makes it a right Γ-module.)

Conversely, given any integral representation of Γ in $GL(n,\mathbb{Z})$ it can be shown there exists a K/k-form of G_1 which gives rise to this representation (in the manner described above). The proof of this converse rests on Proposition 3.1.1 and the one-to-one correspondence between automorphisms of X_1 (the character module of the torus G_1) and the automorphisms of G_1 (see remarks following lemma 2.4.3).

Thus there is a one-to-one correspondence between the category of tori defined over k of dimension n and the category of free Γ-modules of rank n (as asserted in Theorem 2.4.8).

Example 2: k-form of a vector space.

Let V_1 be an n-dimensional vector space defined over the prime field. Then (V,f) is a K/k-form of V_1 if V is a vector space defined over k (see §2.3), and f is a K-isomorphism of V onto V_1. Clearly this implies V has dimension n, and in fact, there is an isomorphism defined over k from V onto V_1. Corollary 3.1.4' is true with V_1 replacing G_1; since there is just one k-isomorphism class of k-forms, namely the class of V_1, we see that $H^1(k,GL(V_1)) = \{1\}$. This is a generalization of Hilbert's Theorem 90. The definitions and theorem can be extended to the case of V_1 a vector space over a division algebra $\mathfrak{K}$; i.e., $GL(n,\mathfrak{K})$ is cohomologically trivial (see [C], Chap. X).

Example 3: k-form of $\mathrm{End}(V_1)$.

Let V_1 be as in Example 2, and $\mathfrak{A}_1 = \mathrm{End}(V_1) = \mathfrak{M}_n$, the total matrix algebra of degree n. If $(\mathfrak{A},f)$ is a k-form of $\mathfrak{A}_1$, then $\mathfrak{A}$ is a normal (or central) simple algebra of degree n, i.e., of dimension n^2 over k, and split by $\bar{k}$.[3)] If (φ_σ) is the system corresponding to $(\mathfrak{A},f)$, then $(\varphi_\sigma) \in H^1(k,\mathrm{Aut}(\mathfrak{M}_n))$. But $\mathrm{Aut}(\mathfrak{M}_n) = PL(n)$ (Skolem-Noether), so taking $\phi_\sigma \in GL(n)$, a representative of the class of φ_σ in $PL(n)$, we see that the system (ϕ_σ) satisfies the property:

$$(\star) \qquad \phi_\sigma{}^\tau\phi_\tau = \lambda_{\sigma,\tau}\phi_{\sigma\tau}, \qquad \lambda_{\sigma,\tau} \in \bar{k}.$$

From this relation, it is easy to verify that the system $(\lambda_{\sigma,\tau})$ is a 2-cocycle of Γ in $\bar{k}^*$, i.e.,

$$(\star\star) \qquad \lambda_{\sigma,\tau}^{\rho}\,\lambda_{\sigma\tau,\rho} = \lambda_{\sigma,\tau\rho}\,\lambda_{\tau,\rho}.$$

The class of the 2-cocycle $(\lambda_{\sigma,\tau})$ in $H^2(\Gamma,\bar{k}^*)$ is uniquely determined by the isomorphism class of $\mathfrak{A}$, i.e., given $(\mathfrak{A}',f')$ another k-form of $\mathfrak{A}_1$, and the analogous systems determined, $(\mathfrak{A}',f') \to (\varphi_\sigma') \to (\phi_\sigma') \to (\lambda_{\sigma,\tau}')$, then $(\mathfrak{A},f)$ and $(\mathfrak{A}',f')$ are isomorphic if and only if $(\lambda_{\sigma,\tau})$ and $(\lambda_{\sigma,\tau}')$ are cohomologous.

If $\mathfrak{A}$ and $\mathfrak{B}$ are both normal simple algebras over k of dimensions n^2 and m^2 respectively, then $\mathfrak{A} \otimes \mathfrak{B}$ is a normal simple algebra of dimension n^2m^2. If $\mathfrak{A} \to (\lambda_{\sigma,\tau})$ and $\mathfrak{B} \to (\mu_{\sigma,\tau})$, 2 cocycles of Γ in $\bar{k}^*$, then $\mathfrak{A} \otimes \mathfrak{B} \to (\lambda_{\sigma,\tau}\mu_{\sigma,\tau})$. (Perhaps we should remark that the algebra $\mathfrak{M}_n \to (1)$, the identity 2-cocycle). If we define an equivalence relation on the set of normal simple algebras by: $\mathfrak{A} \sim \mathfrak{B}$ if and only if $\mathfrak{A} \otimes \mathfrak{M}_r \cong \mathfrak{B} \otimes \mathfrak{M}_s$ for some r, s, then we can define a multiplication on the set of equivalence classes by: $[\mathfrak{A}][\mathfrak{B}] = [\mathfrak{A} \otimes \mathfrak{B}]$. This set of equivalence classes becomes a group with respect to the multiplication defined, called the "Brauer group," $\mathcal{B}(k)$. On the other hand, if $\mathfrak{A} \to (\lambda_{\sigma,\tau})$, then the cohomology class of $(\lambda_{\sigma,\tau})$ is uniquely determined by the class $[\mathfrak{A}]$ of $\mathfrak{A}$, and the mapping $[\mathfrak{A}] \to [\lambda_{\sigma,\tau}]$ defines a homomorphism of $\mathcal{B}(k)$ into $H^2(\Gamma,\bar{k}^*)$. In fact, using the "crossed product," one can see that $\mathcal{B}(k)$ is isomorphic to $H^2(\Gamma,\bar{k}^*)$. (See e.g., [C], or Artin, Nesbitt, Thrall, "Rings with Minimum Condition," Chap. 7,8.)

Example 4: $R_{K/k}(G_1)$. (Weil, Adèles and algebraic groups, Lecture Notes, IAS, Princeton, 1961)

Let G_1 be an algebraic group defined over K, dim $G_1 = n$, and degree $K/k = d$. We wish to prove the existence of an algebraic group, denoted $R_{K/k}(G_1)$, defined over k, and of dimension nd. (It should be noted that $R_{K/k}$ is a functor used in the technique of "descent" of the ground field of definition of various 'algebraic' objects; $R_{K/k}$ maps a category of objects defined over K to a category of objects defined over k.)

Let $\{\sigma_1,\ldots,\sigma_d\}$ be a maximal set of elements in Γ having distinct restrictions on K, and assume σ_1 = identity. Let $\tilde{G}_1 = G_1^{\sigma_1} \times \ldots \times G_1^{\sigma_d}$; $\tilde{G}_1$ is the direct product of all of the distinct conjugates of G_1 by Γ, and $\tilde{G}_1$ is defined over $\bigcup_{i=1}^{d} K^{\sigma_i}$, the smallest Galois extension of k containing K. Let $\Gamma_1 = \mathrm{Gal}(\bar{k}/K)$; then $\Gamma = \bigcup_{i=1}^{d} \Gamma_1\sigma_i$. If $\sigma \in \Gamma$, then the right multiplication of σ permutes the cosets $\Gamma_1\sigma_i$; let σ also denote this permutation, thus $i^\sigma = j$ if and only if $\Gamma_1\sigma_i\sigma = \Gamma_1\sigma_j$. For $\sigma \in \Gamma$, define an isomorphism $\varphi_\sigma\colon \tilde{G}_1 \to \tilde{G}_1^\sigma = G_1^{\sigma_1\sigma} \times \ldots \times G_1^{\sigma_d\sigma}$ by $\varphi_\sigma(g_1,\ldots,g_d) = (g_{1^\sigma},\ldots,g_{d^\sigma})$. It is easily verified that the system $\{\varphi_\sigma\}$ satisfies conditions (1) and (2), so by Prop. 3.1.1, there exists a k-form $(\tilde{G},\tilde{f})$ of $\tilde{G}_1$ corresponding to the system $\{\varphi_\sigma\}$; clearly $\tilde{G}$ is an algebraic group defined over k of dimension nd. Let p_i be the projection of $\tilde{G}_1$ onto its ith factor; it is easily seen that $p_i^\sigma \circ \varphi_\sigma = p_{i^\sigma}$.

If we define $p = p_1 \circ \tilde{f}$, then p is a K-homomorphism of $\tilde{G}$ onto G_1. (To see p is defined over K, note that if $\sigma \in \Gamma_1$, then $1^\sigma = 1$, so $p_1 \circ \varphi_\sigma = p_1$; from this it follows that $p^\sigma = p_1^\sigma \circ \tilde{f}^\sigma = p_1 \circ \varphi_\sigma \circ \tilde{f} = p_1 \circ \tilde{f} = p$ for all $\sigma \in \Gamma_1$.) Also, $p_i \circ \tilde{f} = p^{\sigma_i}$, so that $\tilde{f}$ is uniquely determined by p, $\tilde{f} = p^{\sigma_1} \times \ldots \times p^{\sigma_d}$. (Note $p^{\sigma_i} = p_1^{\sigma_i} \circ \tilde{f}^{\sigma_i} = p_1^{\sigma_i} \circ \varphi_{\sigma_i} \circ \tilde{f} = p_i \circ \tilde{f}$.) We write $(\tilde{G},p)$ for the k-form $(\tilde{G},\tilde{f})$, and define $(\tilde{G},p) = R_{K/k}(G_1)$.

The k-form $R_{K/k}(G_1)$ is unique up to k-isomorphism. We show, in fact, that if a pair $(\tilde{G}',p')$ is given, with p' a K-homomorphism of $\tilde{G}'$ onto G_1 and $\tilde{f}' = p'^{\sigma_1} \times \ldots \times p'^{\sigma_d}$, then the mapping $\psi = \tilde{f}^{-1} \circ \tilde{f}'$ is defined over k. (Thus if $(\tilde{G}',\tilde{f}')$ is a k-form of $\tilde{G}_1$, the mapping ψ is a k-isomorphism of $\tilde{G}'$ onto $\tilde{G}$.)

$$\begin{array}{ccc} \tilde{G} & \xrightarrow{p} & G_1 \\ \psi \uparrow & \nearrow_{p'} & \\ \tilde{G}' & & \end{array}$$

From the definitions of φ_σ, $\tilde{f}$ and $\tilde{f}'$, we have for $\sigma \in \Gamma$, $\tilde{f}^\sigma = \varphi_\sigma \circ \tilde{f}$ and $\tilde{f}'^\sigma = p'^{\sigma_1\sigma} \times \ldots \times p'^{\sigma_d\sigma} = \varphi_\sigma \circ \tilde{f}'$, so that $\psi^\sigma = \tilde{f}^{-\sigma} \circ \tilde{f}'^\sigma = \tilde{f}^{-1} \circ \varphi_\sigma^{-1} \circ \varphi_\sigma \circ \tilde{f}' = \psi$.

Remark: If A_1 is an algebraic set defined over K, then we define $R_{K/k}(A_1)$ to be any pair $(\tilde{A},p)$ where $\tilde{A}$ is an algebraic set defined over k, p/K is a polynomial map of $\tilde{A}$ onto A_1, and $\tilde{f} = p^{\sigma_1} \times \ldots \times p^{\sigma_d}$ is a $\bar{k}$-isomorphism of $\tilde{A}$ onto $\tilde{A}_1 = A_1^{\sigma_1} \times \ldots \times A_1^{\sigma_d}$. In the next section we show Prop. 3.1.1 is true for $G_1 = A_1$ an algebraic set; clearly our arguments in Example 4 then hold for A_1 replacing G_1, and so $R_{K/k}(A_1)$ is unique up to k-isomorphism.

3.3 The proof of Proposition 3.1.1

We prove this proposition in several steps; along the way, we give a direct proof of the existence of $R_{K/k}(A_1)$ and of K/k-forms for algebraic sets.

1° Existence of $R_{K/k}(A_1)$, where A_1/K is an algebraic set.

As in Example 4, let $d = \deg K/k$ and $\{\sigma_1,\ldots,\sigma_d\}$ be a maximal set of elements in Γ with distinct restrictions on K, σ_1 = identity.

Case 1. $A_1 = \Omega$, one-dimensional affine space.

Let $\tilde{A} = \Omega^d$, and let $\{w_1,\ldots,w_d\}$ be a vector-space basis for K over k. Define a polynomial mapping $p: \tilde{A} \to A_1$ by $p(u_1,\ldots,u_d) = \sum_{i=1}^{d} u_i w_i$. Clearly p is defined over k and $p^{\sigma_j}(u_1,\ldots,u_d) = \sum_{i=1}^{d} u_i w_i^{\sigma_j}$; define $\tilde{f}:\Omega^d \to \Omega^d$ by $\tilde{f} = p^{\sigma_1} \times \ldots \times p^{\sigma_d}$. The polynomial map $\tilde{f}$ has matrix $(w_i^{\sigma_j})$ and $\det(w_i^{\sigma_j}) \neq 0$ (since K/k is separable) so f is a $\bar{k}$-isomorphism of algebraic sets. Thus $(\tilde{A},\tilde{f}) = R_{K/k}(A_1)$.

Case 2. Let A_1, B_1 be algebraic sets defined over K, and suppose that $R_{K/k}(A_1)$ and $R_{K/k}(B_1)$ exist. Then $R_{K/k}(A_1 \times B_1)$ exists and is gotten by taking direct products of 'everything'. The verification is left as an exercise.

Case 3. Let A_1 and B_1 be algebraic sets defined over K, with $B_1 \subset A_1$, and suppose $R_{K/k}(A_1) = (\tilde{A},\tilde{f})$ exists. We show $R_{K/k}(B_1)$ also exists. Since $\tilde{f}$ is a $\bar{k}$-isomorphism of $\tilde{A}$ onto $\tilde{A}_1 = A_1^{\sigma_1} \times \ldots \times A_1^{\sigma_d}$ and $\tilde{B}_1 = B_1^{\sigma_1} \times \ldots \times B_1^{\sigma_d}$ is an algebraic subset of $\tilde{A}_1$, it follows that $\tilde{B} = \tilde{f}^{-1}(\tilde{B}_1)$ is an algebraic subset of $\tilde{A}$. To show $R_{K/k}(B_1) = (\tilde{B},\tilde{f}|\tilde{B})$, we only need to show $\tilde{B}$ is defined over k. Now $\tilde{B}_1$ is defined over $\bigcup_{i=1}^{d} K^{\sigma_i}$, and $\tilde{f}^{-1}/\bar{k}$, so $\tilde{B} = \tilde{f}^{-1}(\tilde{B}_1)$ is defined over $\bar{k}$. Also, for $\sigma \in \Gamma$, $\tilde{B}^{\sigma} = \tilde{f}^{-\sigma}(\tilde{B}_1^{\sigma}) = \tilde{f}^{-1} \circ \varphi_{\sigma}^{-1}(\tilde{B}_1^{\sigma}) = \tilde{f}^{-1}(\tilde{B}_1) = \tilde{B}$ (since B_1/K, one has $\tilde{B}_1^{\sigma} = \varphi_{\sigma}(\tilde{B}_1)$) so $\tilde{B}/k$ (Corollary, Prop. 1.1.1).

Since every algebraic set can be considered as a subset of Ω^n for some n, the three cases together imply the existence of $R_{K/k}(A_1)$ for any algebraic set A_1/K.

2° Universal properties of $R_{K/k}(A_1)$

In Example 4, (§3.2) and the following Remark, we have noted the uniqueness of $R_{K/k}(A_1)$; we also have the following 'universality' conditions, which are easily verified. Let $(\tilde{A},\tilde{f}) = R_{K/k}(A_1)$, and $p = p_1 \circ \tilde{f}$. If $\tilde{B}/k$ is an algebraic set, and ϕ is a polynomial mapping of $\tilde{B}$ into A_1 defined over K, then there exists a polynomial mapping ψ/k of $\tilde{B}$ into $\tilde{A}$ so that the following diagram commutes:

$$\begin{array}{ccc} \tilde{A} & \xrightarrow{p} & A_1 \\ \psi \uparrow & \nearrow_{\phi} & \\ \tilde{B} & & \end{array}$$

If B_1 is an algebraic set defined over K, $(\tilde{B},\tilde{g}) = R_{K/k}(B_1)$, $q = p_1 \circ \tilde{g}$, and ψ/K a polynomial mapping of A_1 into B_1, then there exists a unique polynomial mapping $\tilde{\psi}/k$ which makes the following diagram commute: ($\tilde{\psi}$ is denoted $R_{K/k}(\psi)$)

$$\begin{array}{ccc} \tilde{A}/k & \xrightarrow{p/K} & A_1/K \\ \tilde{\psi}/k \downarrow & & \downarrow \psi/K \\ \tilde{B}/k & \xrightarrow{q/K} & B_1/K \end{array}$$

3° If $A_1 = G_1$ is an algebraic group defined over K, then $R_{K/k}(G_1)$ is a group, and p is a homomorphism. This fact is proved using the properties in 2°. If $(\tilde{G},p) = R_{K/k}(G_1)$, then $(\tilde{G}\times\tilde{G}, p\times p) = R_{K/k}(G_1 \times G_1)$; let $\psi: G_1 \times G_1 \to G_1$ be the mapping of group multiplication in G_1. By 2°, there exists a polynomial mapping $\tilde{\psi}/k$ making the diagram below commute; clearly $\tilde{\psi}$ defines a multiplication in $\tilde{G}$.

$$\begin{array}{ccc} \tilde{G}\times\tilde{G} & \xrightarrow{p\times p} & G_1 \times G_1 \\ \downarrow \tilde{\psi} & & \downarrow \psi \\ \tilde{G} & \xrightarrow{p} & G_1 \end{array}$$

Similarly, if $\psi':G_1 \to G_1$ maps $g_1 \to g_1^{-1}$ for $g_1 \in G_1$, then the polynomial mapping $\tilde{\psi}'/k$ induced by ψ defines an inverse operation in $\tilde{G}$.

$$\begin{array}{ccc} \tilde{G} & \xrightarrow{p} & G_1 \\ \downarrow{\tilde{\psi}'} & & \downarrow{\psi'} \\ \tilde{G} & \xrightarrow{p} & G_1 \end{array}$$

It is also clear that with respect to this group structure on $\tilde{G}$ that p is a homomorphism of $\tilde{G}$ onto G_1.

4° Proof of Prop. 3.1.1 for $G_1 = A_1$ any algebraic set.

We are given A_1/K, and a system (φ_σ) of rational isomorphisms $\varphi_\sigma : A_1 \to A_1^\sigma$, satisfying (1) and (2). With notations as in 1° and 2°, let $(\tilde{A},\tilde{f}) = R_{K/k}(A_1)$, and define

$$A = \{x \in \tilde{A} \mid p^\sigma(x) = \varphi_\sigma \circ p(x) \quad \text{for all } \sigma \in \Gamma\}.$$

Condition (2) implies that $x \in A$ if and only if $p^{\sigma_i}(x) = \varphi_{\sigma_i} \circ p(x)$, $1 \leq i \leq d$, thus $x \to p(x)$ is an injection of A into A_1. But also, if $y \in A_1$, and we put $x = \tilde{f}^{-1}(\varphi_{\sigma_1}(y),\ldots,\varphi_{\sigma_i}(y),\ldots,\varphi_{\sigma_d}(y))$, then $x \in A$ and $y = p(x)$. Thus $p|A$ is an isomorphism of A onto A_1. By its definition, it is clear that A is an algebraic subset of $\tilde{A}$, and defined over $\bar{k}$, but condition (1) implies $A^\sigma = A$ for all $\sigma \in \Gamma$, so A is defined over k (Corollary to Prop. 1.1.1). If we let $f = p|A$, then (A,f) is a K/k-form of A_1 with corresponding system (φ_σ). If $A_1 = G_1$ is a group, then it is easily verified that A is a subgroup of the group $R_{K/k}(G_1)$, and (by 3°) f is an isomorphism of algebraic groups.

<u>Remark</u>: The k-rational points of $R_{K/k}(A_1)$ can be identified with the K-rational points of A_1 via the map p. (This follows from Case 3, taking B_1 to be the set consisting of just one K-rational point.)

<u>Exercise 1</u>. Let $\tilde{G} = R_{K/k}(\mathbb{G}_m)$. Show that the character module $\tilde{X}$ of $\tilde{G}$ is

isomorphic to the Γ-module generated by the cosets of $\Gamma_1\backslash\Gamma$; or in other words, that the corresponding integral representation of Γ is the "monomial representation" defined by $\Gamma_1\backslash\Gamma$.

Exercise 2. How many indecomposable tori are there over $\mathbb{R}$? (Count up all indecomposable integral representations of $\mathrm{Gal}(\mathbb{C}/\mathbb{R}) = \{1,\sigma_0\}$.)

§4. The structure of algebraic groups

4.1 Semi-simple and unipotent elements ([1]-II,III; [2] exposé 4,6)

Definition: An element $g \in GL(V)$ is called semi-simple (resp. unipotent) if and only if g is similar to a diagonal matrix (resp., $(g-1)^m = 0$ for some integer m).

For example, any 'diagonalizable' matrix in $GL(n)$ is semi-simple, and an upper (or lower) triangular matrix having all 1's on the diagonal is unipotent.

Proposition 4.1.1: Every element $g \in GL(V)$ has a unique decomposition $g = g'g''$ where g' is semi-simple, g'' is unipotent, and $g'g'' = g''g'$.

This result can be gotten from the Jordan normal form of g; in fact, g' and g'' are each polynomials of g.

Notation: We will denote the elements g', g'' in Prop. 4.1.1 by g_s and g_u, respectively.

Proposition 4.1.2: Let G be an algebraic group.

(1) If $g \in G$, then g_s and g_u are in G, and g_s and g_u are uniquely determined by the structure of G (i.e., they are independent of the representation of G in some $GL(V)$).

(2) If G' is an algebraic group and φ a rational homomorphism $\varphi: G \to G'$,

then $\varphi(g_s) = \varphi(g)_s$, $\varphi(g_u) = \varphi(g)_u$ for all $g \in G$.

The first part of Prop. 4.1.2 says that we can speak without ambiguity of the semi-simple and unipotent elements of an algebraic group G. Thus, we denote $G_s = \{g \in G | g = g_s\}$ and $G_u = \{g \in G | g = g_u\}$. If $G = G_u$, we say G is a unipotent group.

Example 1: Let G be a torus, $G \simeq D(n)$. Clearly every element of G is semi-simple. Conversely, it can be shown that if $G = G_s$, and G is connected, then G is a torus. ([2])

Example 2. Let $G = \mathbb{G}_a$. Since $\mathbb{G}_a$ has a representation in GL(2) by $\xi \to \begin{pmatrix} 1 & \xi \\ 0 & 1 \end{pmatrix}$, we see that $\mathbb{G}_a$ is unipotent. Conversely, it can be shown that $\mathbb{G}_a$ is the unique one-dimensional connected unipotent algebraic group ([2], 7-06, Theorem 4).

Proposition 4.1.3: (1) If G is a connected unipotent algebraic group, then G is nilpotent (both as an abstract and as an algebraic group); that is, there exists a finite series $\{G_i\}$ of closed, connected normal subgroups of G satisfying $G = G_0 \supset G_1 \supset \ldots \supset G_r = \{1\}$, $[G, G_i] \subset G_{i+1}$; moreover, we may take G_i's in such a way that $G_i/G_{i+1} \overset{\varphi_i}{\simeq} \mathbb{G}_a$ (φ_i a rational isomorphism). If G is defined over k, and k is perfect, then the groups G_i and isomorphisms φ_i can all be taken to be defined over k.
(2) If G/k is commutative and unipotent, and k has characteristic zero, then G is isomorphic to $\mathbb{G}_a{}^n$ for some integer n (hence G is connected). In the characteristic p case G is not necessarily connected nor isomorphic to $\mathbb{G}_a{}^n$ (see [2], 4-12).

4.2 Solvable groups ([1]-III; [2] exposé 6; [13]-III)

Definition: An algebraic group G is solvable if it is solvable as an abstract group.

Since the "commutator group" $[H_1, H_2]$ of two closed subgroups H_1, H_2 of an algebraic group is closed, it is easy to show that the 'composition' series involved in the definition of solvability can be taken to consist of closed subgroups of G. ([1],3-03)

<u>Proposition 4.2.1</u>: Let G be a connected solvable algebraic group defined over k.

(1) There exists an isomorphism of G into Tr(n), the group of all upper triangular matrices of degree n (Lie-Kolchin).

(2) G_u is a k-closed connected normal subgroup of G.

(3) There exists a maximal torus T (defined over k) of G such that $G = T \cdot G_u$ (semi-direct product).

(4) All maximal tori of G are conjugate by inner automorphisms of G.

<u>Remark 1</u>. The product $G = T \cdot G_u$ in (3) is semi-direct in the sense of algebraic groups, namely, a semi-direct product of abstract groups which is a direct product of algebraic sets.

<u>Remark 2</u>. The proof of the existence of a maximal torus T defined over k is found in Theorem 4, [7]. In fact, if G is <u>any</u> connected algebraic group defined over k, then G contains a maximal torus T defined over k. (See [6], p. 45 for k perfect; see [A], p. 26 for the general case.)

<u>Definition</u>: Let G be a connected algebraic group. The <u>radical of</u> G, denoted R, is a maximal element in the set of connected solvable normal subgroups of G. The <u>unipotent radical of</u> G, R_u, is the unipotent part of R. G is <u>semi-simple</u> if and only if $R = \{1\}$. G is <u>reductive</u> if and only if $R_u = \{1\}$.

The radical R of G is uniquely determined, for any two connected solvable normal subgroups of G generate a connected solvable normal subgroup. It is easily verified that G/R and G/R_u are semi-simple and reductive, respectively.

In classifying semi-simple (or reductive) algebraic groups, the next result allows us to reduce essentially to the case of simple groups.[4)]

Proposition 4.2.2: Let G be a connected algebraic group.

(1) If G is semi-simple, then G is isogeneous to a direct product of simple groups.

(2) If G is reductive, then G is isogeneous to the direct product of a semi-simple group with a torus. (See [2], exposé 17)

If the characteristic of the field k is zero, then any connected algebraic group G defined over k can be decomposed into the semi-direct product of a reductive subgroup of G with the unipotent radical of G. (In characteristic zero, there is a one-to-one correspondence between algebraic groups and their Lie algebras, so that structure theorems for Lie algebras give rise to corresponding structure theorems for algebraic groups.)

4.3 Borel subgroups ([1]-IV; [2] exposés 6,7,9,10; [13]-IV)

Throughout this section, G is a connected algebraic group defined over k.

Definition: A Borel subgroup of G is a maximal closed connected solvable subgroup of G.

Proposition 4.3.1:

(1) All Borel subgroups of G are conjugate to each other by inner automorphisms of G.

(2) If B is a Borel subgroup of G, then the coset space G/B is a projec-

tive (hence complete) algebraic variety.

(3) Every Borel subgroup is its own normalizer in G.

(4) If B is a Borel subgroup of G, then $G = \bigcup_{g \in G} g B g^{-1}$.

An essential step in the proof of the properties (1),(2) is given by the following fixed point theorem.

Proposition 4.3.2: (Borel)

Let G be a connected solvable algebraic group which acts on a complete algebraic variety V, i.e., there is a polynomial mapping $G \times V \to V$.
$(g,v) \to gv$.

Then there exists a point $v \in V$ fixed by G, that is, $gv = v$ for all $g \in G$.

(See [1] Chap. IV; [2], 5-14).

To illustrate how to use Proposition 4.3.2, we consider the special case where G = GL(n), and prove the properties (1),(2) (in other words, we shall prove the Lie-Kolchin theorem).

Example: The Flag Manifold

Let G = GL(n), and $B_o = Tr(n)$, the subgroup of upper triangular matrices. The coset space G/B_o is identified with the projective variety $\mathcal{F}$ called the "Flag Manifold." We now describe $\mathcal{F}$.

Let V be an n-dimensional vector space, with basis $(e_1,\ldots,e_n)$, and fix W_o the r-dimensional subspace of V spanned by $e_1,\ldots,e_r$. Let $\mathcal{G}_{n,r}$ be the set of all r-dimensional subspaces of V; G acts on $\mathcal{G}_{n,r}$ by $W \to gW$, and this action is transitive. The stabilizer of W_o under this action of G is just the subgroup H of matrices of the type

$$\begin{array}{c} \\ r \end{array}\!\!\begin{pmatrix} \overset{r}{*} & \overset{n-r}{*} \\ 0 & * \end{pmatrix}.$$

Thus the coset space G/H is identified to $\mathcal{G}_{n,r}$ in a natural way. The set $\mathcal{G}_{n,r}$ can be given the structure of a projective variety in the following manner. For each $W \in \mathcal{G}_{n,r}$, fix a basis $x_1, \ldots, x_r$ of W, and form the exterior product $x_1 \wedge \ldots \wedge x_r$. This exterior product is uniquely determined by W up to a scalar multiple. Thus there is a natural embedding of $\mathcal{G}_{n,r}$ in $P(\Lambda^r(V))$ given by $W \to \langle x_1 \wedge \ldots \wedge x_r \rangle$. The set $\mathcal{G}_{n,r}$ considered as a subset of $P(\Lambda^r(V))$ is characterized by certain polynomial relations (Plücker's relations), so that it is an algebraic subvariety of $P(\Lambda^r(V))$. The set $\mathcal{G}_{n,r}$ with this structure is called the "Grassmann Manifold."

Now let $\mathcal{F}$ be the set of all series of subspaces of V, $(V_0, V_1, V_2, \ldots, V_n)$ with $V_0 = V \supset V_1 \supset \ldots \supset V_n = \{0\}$, and $\dim V_i = n - i$ (the series $(V_0, \ldots, V_n)$ is called a <u>flag</u>). Then G acts transitively on $\mathcal{F}$ by $(V_0, \ldots, V_n) \to (gV_0, \ldots, gV_n)$, and B_0 is just the stabilizer of the flag $(V, W_1, \ldots, W_n)$ with W_i spanned by $e_1, \ldots, e_{n-i}$. Thus G/B_0 is identified to $\mathcal{F}$ in a natural way. The set $\mathcal{F}$, which is a subset of the 'direct product' of the $\mathcal{G}_{n,r}$ has the structure of a projective variety since it is known that the direct product of projective spaces can be considered as a subvariety of a projective space, and also, the set $\mathcal{F}$ of flags can be characterized by polynomial relations. The set $\mathcal{F}$, with its structure as a projective variety, is called the Flag Manifold.

In order to obtain the Lie-Kolchin theorem, we now apply Prop. 4.3.2 to $\mathcal{F}$. If $G_1 \subset GL(V)$ is solvable and connected, then G_1 acts on the flag manifold $\mathcal{F}$ of V, and since $\mathcal{F}$ is complete, G_1 leaves fixed a flag $(V, V_1, \ldots, V_n) \in \mathcal{F}$. Thus there is a basis of V such that with respect to it, G_1 has triangular form. This shows that any such group G_1 is conjugate to a subgroup of $B_0 = Tr(n)$, hence B_0 is a Borel subgroup of $G = GL(n)$. The properties (1) and (2) follow immediately.

Remarks: 1. From Prop. 4.2.1, (4) and Prop. 4.3.1,(1) it follows that all maximal tori of G are conjugate.

2. Prop. 4.3.1,(2) can be made more precise: Let H be any closed connected subgroup of G. G/H is a complete variety if and only if H contains a Borel subgroup of G.

Definition: Any subgroup of G which contains a Borel subgroup is called a parabolic subgroup of G.

Later we will study in detail the structure and properties of parabolic subgroups of a reductive group G. It should be noted that one property of parabolic subgroups which follows from Prop. 4.3.1 (3) is that any parabolic subgroup is its own normalizer, hence any parabolic subgroup is connected. (See [3], Sec. 4.)

4.4 Questions of rationality ([3], [6], [7], [12], [13]-V)

In this section we concern ourselves with "relativizing" the results of sections 4.1, 4.2, 4.3 taking into account the field of definition (of groups, isomorphisms, vector spaces, etc.). Throughout, G is a connected algebraic group.

Proposition 4.4.1: The following three conditions are equivalent:

1) G is solvable, defined over k, and all (rational) characters of G are defined over k.
2) There exists a k-isomorphism of G into Tr(n).
3) G = T·U, a semi-direct product over k, where T is a k-trivial torus of G, and U is a unipotent subgroup of G defined over k.

The proof of this proposition can be found in [3],[6],[7]; however, we indicate an outline of the argument. First, we need to

Remark: There are no non-trivial characters of a unipotent group. For, if χ is a character of U, a unipotent group, then $\chi(g) \in \mathbb{G}_m$ for $g \in U$. Since g is unipotent, $\chi(g)$ is unipotent (Prop. 4.1.2,(2)), but every element of $\mathbb{G}_m$ is semi-simple, thus $\chi(g) = 1$.

Proof of 4.4.1: 1) => 2). Let $G \subset GL(V)$; it suffices to show there is a flag $(V, V_1, \ldots, V_n) \in \mathcal{F}$ which is defined over k (i.e., each V_i/k) and fixed by G. By using induction on dim V, it suffices to show G fixes a one-dimensional subspace of V defined over k. Let $\chi \in X(G)$, and define $V_\chi = \{v \in V \mid gv = \chi(g)v, \text{ all } g \in G\}$. By the Lie-Kolchin theorem, there exists a χ such that $V_\chi \neq \{0\}$; since χ is defined over k, V_χ is defined over k. Take $e_1 \in (V_\chi)_k$, and extend to a basis $(e_1, \ldots, e_n)$ of V, then with respect to this basis, elements of G have the form $n-1\left\{\begin{pmatrix} * & * \\ 0 & * \end{pmatrix}\right.$.

2) => 3). If k is perfect, this is easily proved; for $G \subset Tr(n)$ implies $G_u = G \cap Tr(n)_u$ is defined over k, and there exists a maximal torus T/k such that $G = T \cdot G_u$ (Prop. 4.2.1). Since $T \cong G/G_u \subset Tr(n)/Tr(n)_u \simeq D(n)$, (these isomorphisms and embedding are defined over k), and D(n) splits over the prime field, T splits over k. If k is not perfect, then (since the intersection of two algebraic sets defined over k is not necessarily defined over k), a strong argument is needed to prove G_u/k. One way to prove this is to notice that G_u is the kernel of the separable k-homomorphism which maps elements of G onto their diagonal parts in Tr(n).

3) => 1). $G = T \cdot U$ implies $X(G) = X(T)$ (see Remark above), and T k-trivial implies all elements of X(G) are defined over k. (Prop.2.4.6).

Definition: A group G which satisfies any of the three equivalent conditions of Prop. 4.4.1 is called k-solvable.

If k is perfect, then a k-solvable group G has a composition series $G = G_o \supset G_1 \supset \ldots \supset G_r = \{1\}$, G_i/G_{i+1} isomorphic to $\mathbb{G}_a$ or $\mathbb{G}_m$, all groups and isomorphisms defined over k.

Definition: A k-Borel subgroup of G is a maximal connected k-solvable subgroup.

Proposition 4.4.2: If G is defined over k, then all k-Borel subgroups of G are conjugate by k-rational inner automorphisms of G; that is, if B_1 and B_2 are k-Borel subgroups of G, then there exists an element $g \in G_k$ such that $gB_1g^{-1} = B_2$.

Corollary 4.4.2: If G is defined over k, then all maximal k-trivial tori of G are conjugate with respect to k-rational inner automorphisms of G.

Proposition 4.4.2 and its corollary can be proved in a manner similar to that of Prop. 4.3.1, using the following revised 'fixed point theorem' (4.3.2).

Proposition 4.4.3: Let G be a connected k-solvable group, and V a complete variety defined over k satisfying $V_k \neq \emptyset$, and suppose G operates on V over k ($G \times V \rightarrow V$ is a k-rational polynomial map). Then there exists a point $v \in V_k$ which is fixed by G.

Definition: The k-rank of G/k is the dimension of any maximal k-trivial torus. The rank of G is the Ω-rank of G.

A group G/k is called k-compact or k-anisotropic, if G has no non-trivial k-Borel subgroup (i.e., the only k-Borel subgroup of G is $\{1\}$).

The term k-compact is used in this definition since this is the analogue of G being compact, if G is a real Lie group. Later we will prove (for k a perfect field) that when G/k is reductive, G is k-compact if and only if the k-rank of G is zero. When G/k is semi-simple, G is k-compact if and only if G has no unipotent subgroup defined over k.

Although in general G/H is not a complete variety when H is a k-Borel subgroup of G, we have the following analogue to Prop. 4.3.1,(2).

Proposition 4.4.4: Let G/k, and H a k-Borel subgroup of G. There exists a complete variety V defined over k on which G operates, and an injective polynomial map $G/H \to V$ preserving multiplication by G such that G_k acts transitively on V_k.

Thus one can identify the coset space G_k/H_k with the set V_k of k-rational points of a complete variety V.

If k is a local field, say $\mathbb{R}$ or $\mathbb{Q}_p$, then it is known that V_k is compact (in the 'usual' topology on k). In this case, if G is k-compact, then $H = \{1\}$, so $G_k = V_k$ is compact. The converse also holds, so we have

Corollary 4.4.4: Let k be a local field, and G/k. G is k-compact if and only if G_k is compact.

Example 1 (char. k $\neq 2$): Let S be a non-degenerate symmetric bilinear form defined over k, and let $G = SO(V,S)$. (See Example 3, 2.3.) We show that the k-rank of G is equal to the k-Witt index of S, thus G is k-compact if and only if S does not express zero non-trivially.

We first recall the definition of k-Witt index of S. V_k can be decomposed as $V_k = \sum_{i=1}^{r}(ke_i + ke_i') + V_{ok}$ where $S(e_i,e_i') = 1$, $1 \leq i \leq r$, and for $x \in V_{ok}$, $S(x,x) = 0$ implies $x = 0$. With respect to this decomposition of V_k, S has matrix $\begin{pmatrix} \begin{matrix} 0 & 1_r \\ 1_r & 0 \end{matrix} & 0 \\ 0 & S_o \end{pmatrix}$, and S_o does not express zero non-trivially. The k-Witt index of V is r, and this is uniquely determined.

1°. k-rank G $\geq$ r.

The set of matrices of the type

$$\begin{pmatrix} \begin{matrix} \xi_1 & & 0 \\ & \ddots & \\ 0 & & \xi_r \end{matrix} & 0 & \\ 0 & \begin{matrix} \xi_1^{-1} & & 0 \\ & \ddots & \\ 0 & & \xi_r^{-1} \end{matrix} & 0 \\ & 0 & 1_{n-2r} \end{pmatrix}$$

is clearly a subgroup of G; if we call this subgroup A_1, then A_1 is a k-trivial torus of dimension r. Since the k-rank of G is the dimension of a maximal k-trivial torus of G, the assertion follows.

2°. Suppose $r = 0$ (S is called k-anisotropic); we show this implies G is k-compact. Let H be any k-Borel subgroup of G, and let P(V) be the projective space associated with V. The space P(V) is a complete variety, defined over k, satisfying $P(V)_k \neq \emptyset$, and the operation of H on P(V) is defined over k, so by Prop. 4.4.3, there is a k-rational point $\langle v \rangle \in P(V)$ fixed by H. The representative v of $\langle v \rangle$ can be taken in V_k. Since $r = 0$, $S(v,v) \neq 0$; let $V_1 = \{v\}^\perp$. The space V_1 has dimension $n-1$ and is defined over k, and $V = \{v\} + V_1$. Let $G_1 = \{g \in G | g\{v\} = \{v\}\}$, $G_1 \supset H$. If $g \in G_1$, then $gv = \chi(g)v$, where χ is a character of G; but the orthogonal group has no non-trivial characters, so $gv = v$. Thus elements of G_1 have the form $\left(\begin{array}{c|c} 1 & 0 \\ \hline 0 & g_1 \end{array}\right)$ with $g_1 \in SO(V_1,S_1)$, where $S_1 = S|V_1$; in fact, $G_1 \simeq SO(V_1,S_1)$. By induction on $n = \dim V$, it can be shown that $H = \{1\}$. (It should be noted that when $n = 2$ a different argument than the one given above is needed to prove the character χ is trivial.)

3°. The torus A_1 (of 1°) is a maximal k-trivial torus of G (hence r = k-rank G). Suppose A is a maximal k-trivial torus of G; we can assume $A \supset A_1$. Since A is a commutative group, any element $g \in A$ commutes with any element of A_1, so that g must have the form

$$g = \left(\begin{array}{cc|c} \begin{matrix} \eta_1 & & 0 \\ & \ddots & \\ 0 & & \eta_r \end{matrix} & 0 & \\ 0 & \begin{matrix} \eta_{r+1} & & 0 \\ & \ddots & \\ 0 & & \eta_{2r} \end{matrix} & 0 \\ \hline 0 & & g_0 \end{array}\right).$$

Also, $g \in G$ implies $S(ge_i, ge_i') = S(\eta_i, \eta_{r+i}) = 1$, so that $\eta_{r+i} = \eta_i^{-1}$. Finally, the group of elements $\{g_0 | g \in A\}$ is a k-trivial torus contained in the group $SO(V_0, S_0)$, which by 2°, is a k-compact group. Thus $g_0 = 1$, and we have that $A = A_1$.

<u>Example 2</u>. Let T be a torus defined over k, and split over K, where K/k is a finite Galois extension. Let $\Gamma = Gal(K/k)$. We show that there exist subtori A, T_0 of T satisfying: A is the largest k-trivial subtorus of T, T_0 is the largest k-compact subtorus of T, $T = A \cdot T_0$, and $A \cap T_0$ is finite (in other words, T is isogeneous to $A \times T_0$). Define

$$X^\Gamma = \{\chi \in X \mid \chi^\sigma = \chi, \text{ all } \sigma \in \Gamma\},$$

$$X_0 = \{\chi \in X \mid \sum_{\sigma \in \Gamma} \chi^\sigma = 0\}.$$

These submodules of X are both cotorsion free (i.e., X/X^Γ and X/X_0 are torsion free) and Γ-invariant. Let $T_0 = (X^\Gamma)^\perp$, $A = (X_0)^\perp$; T_0 and A are subtori of T defined over k (Prop. 2.4.1, 2.4.7).

If T' is any subtorus of T defined over k, and $X_1 = (T')^\perp$, then T' is k-trivial if and only if Γ operates trivially on $X(T') = X/X_1$ (this means $\chi^\sigma - \chi \in X_1$ for all $\chi \in X$, $\sigma \in \Gamma$). But this is true if and only if $X_1 \supset X_0$ (note $\chi \in X_0$ implies $\chi = \frac{1}{d}\left(\sum_{\sigma \in \Gamma} (\chi - \chi^\sigma)\right)$ where $d = [\Gamma:1]$) which is equivalent to $T' \subset A$. Thus A is the largest k-trivial subtorus of T. From this it follows that T is k-compact if and only if $A = \{1\}$, or equivalently, $X_0 = X$. Applying this criterion to a subtorus T'/k of T, we see that T' is k-compact if and only if $X(T')_0 = X(T')$ (that is, $\sum_{\sigma \in \Gamma} \chi^\sigma \in X_1$ for all $\chi \in X$). This is true if and only if $X_1 \supset X^\Gamma$ (note $\chi \in X^\Gamma$ implies $\chi = \frac{1}{d}\left(\sum_{\sigma \in \Gamma} \chi^\sigma\right)$), which is equivalent to $T' \subset T_0$. Thus T_0 is the largest k-compact subtorus of T.

Since for any $\chi \in X$, one has $\chi = \frac{1}{d}\left(\sum_{\sigma \in \Gamma} (\chi - \chi^\sigma)\right) + \frac{1}{d}\left(\sum_{\sigma \in \Gamma} \chi^\sigma\right)$, it follows that $X_{\mathbb{Q}} = (X_0)_{\mathbb{Q}} + (X^\Gamma)_{\mathbb{Q}}$, a direct sum. From this decomposition, it is clear that $[X: X_0 + X^\Gamma] < \infty$, and $X_0 \cap X^\Gamma = \{0\}$, hence $A \cap T_0$ is finite, and $T = A \cdot T_0$.

II. GENERAL PRINCIPLES OF CLASSIFICATION

§1. Structure of semi-simple algebraic groups ([2])

1.1 The root system and the Weyl group ([2], 11 ~14; [13]-IV, §14; [E]-VI)

Throughout this section, G is a connected semi-simple algebraic group, and T is a maximal torus of G.

Definition: A character $\alpha \in X = X(T)$ is called a root (of G with respect to T) if there exists an isomorphism x_α of $\mathbb{G}_a$ onto a closed subgroup P_α of G which satisfies

$$(1) \qquad t\, x_\alpha(\xi) t^{-1} = x_\alpha(\alpha(t)\xi) \quad \text{for all } t \in T,\ \xi \in \mathbb{G}_a.$$

The set of all roots of G with respect to T will be denoted $\surd$; it is called the root system of G relative to T.

It is known that the subgroup P_α of G is uniquely determined by the root α ([2]-18-03); from this it follows that the isomorphism x_α is unique up to a scalar multiplication in $\mathbb{G}_a$ (for the only birational automorphisms of $\mathbb{G}_a$ are scalar multiplications). It is also known that the root system $\surd$ of G is a finite set.

If we denote by N(T) and Z(T) the normalizer and centralizer of T in G respectively, then it is known that $Z(T) = T$, and $N(T)/T$ is a finite group. (In general, for any connected algebraic group G, Z(T) is the connected component of N(T) (see [2]).) This group $W = N(T)/T$ is called the Weyl group of G relative to T. W can be regarded as an automorphism group of T, or of X, or of $\hat{X} = \mathrm{Hom}(X,\mathbb{Z})$ in a natural way, namely, to each $s \in N(T)$, we associate an automorphism w_s of T defined by

$$(2) \qquad w_s(t) = s\,t\,s^{-1} \qquad \text{for } t \in T$$

and denote by the same letter w_s the automorphism of X defined by

(3) $\quad w_s(\chi)(w_s(t)) = \chi(t) \qquad$ for $t \in T$, $\chi \in X$

(the automorphism of $\hat{X}$ associated to s is the contragedient of the automorphism w_s of X). The correspondence $s \to w_s$ gives the identification mentioned above. It follows from (1), (2), and (3) that if $\alpha \in \Phi$ and $s \in N(T)$, then $w_s(\alpha) \in \Phi$, and

(4) $\quad s x_\alpha(\xi) s^{-1} = x_{w_s(\alpha)}(\lambda\xi), \qquad \xi \in \mathbb{G}_a,$

for some $\lambda \in \mathbb{G}_m$.

The triple (X, Φ, W) possesses the following list of properties:

$$\begin{cases} X \text{ is a free module of rank } \ell \qquad (\ell = \dim T) \\ \Phi \text{ is a finite subset of } X \\ W \text{ is a finite automorphism group of } X \end{cases}$$

and

(i) $0 \notin \Phi$; if $\alpha \in \Phi$, then $-\alpha \in \Phi$;

(i)* If $\alpha \in \Phi$ and $c\alpha \in \Phi$ for $c \in \mathbb{Q}$, then $c - \pm 1$;

(ii) To each $\alpha \in \Phi$, there corresponds an element $w_\alpha \in W$ which acts as follows:

(5) $\quad w_\alpha(\chi) = \chi - \alpha^*(\chi)\alpha \qquad$ for $\chi \in X$,

where $\alpha^* \in \hat{X}$. Also, $w_\alpha(\Phi) = \Phi$.

(iii) $X_{\mathbb{Q}}$ $(= X \otimes_{\mathbb{Z}} \mathbb{Q})$ is generated by Φ as a linear space over $\mathbb{Q}$.

(iv) W is generated by $\{w_\alpha \mid \alpha \in \Phi\}$.

In general, if Φ is a finite subset of any free module X (of finite rank) and Φ satisfies properties (i), (i)*, (ii) (with $w_\alpha \in \mathrm{Aut}(X)$), and (iii), then Φ is called an (abstract) root system in X. The group W generated by the set $\{w_\alpha \mid \alpha \in \Phi\}$ is finite (by (ii) and (iii)), and is uniquely determined by the pair (X, Φ); W is called the Weyl group of Φ.

For instance, given a root system $\mathcal{V}$ in X, the set $\mathcal{V}^* = \{\alpha^* \mid \alpha \in \mathcal{V}\}$ (which is also uniquely determined by $\mathcal{V}$) is a root system in $\hat{X}$ (the element $\alpha^* \in \mathcal{V}^*$ corresponding to $\alpha \in \mathcal{V}$ in property (ii) is called a "co-root").

From this list of properties of a root system $\mathcal{V}$ we can easily deduce other properties. Since W is finite, one can introduce a W-invariant metric (= positive definite symmetric bilinear form) $\langle \ , \ \rangle$ in $X_{\mathbb{Q}}$. From the relation $\langle w_\alpha(\lambda), w_\alpha(\lambda)\rangle = \langle \lambda, \lambda \rangle$, and (5), one has

$$\alpha^*(\lambda) = \frac{2\langle \alpha, \lambda \rangle}{\langle \alpha, \alpha \rangle}.$$

Thus α^* may be identified with $2\alpha/\langle \alpha, \alpha \rangle$, and it follows that $w_\alpha^2 = 1$. (Since $w_\alpha(\alpha) = -\alpha$, and w_α leaves fixed the hyperplane $\{\lambda \mid \langle \alpha, \lambda \rangle = 0\}$, w_α is called a "reflection," or "symmetry" with respect to α). The numbers $c_{\alpha,\beta} = \frac{2\langle \alpha, \beta \rangle}{\langle \alpha, \alpha \rangle}$, $\alpha, \beta \in \mathcal{V}$ are called Cartan integers, and property (ii) implies $c_{\alpha,\beta} \in \mathbb{Z}$ (this condition is sometimes called the "integrality condition" on a root system). Actually, property (ii) implies much more; we have the following inclusion:

$$(6) \qquad \{\mathcal{V}\}_{\mathbb{Z}} \subset X \subset \{\mathcal{V}^*\}_{\mathbb{Z}}^{\wedge},$$

where $\{\mathcal{V}^*\}^{\wedge} = \{\lambda \in X_Q \mid \langle \alpha^*, \lambda \rangle \in \mathbb{Z} \text{ for all } \alpha^* \in \mathcal{V}^*\}$.

Remark: A finite subset $\mathcal{V}$ of a free module X of finite rank which satisfies only conditions (i),(ii),(iii) is called a "root system in a wider sense." Although (i)* cannot be expected to be satisfied by such a root system, the integrality condition implies that if $\alpha \in \mathcal{V}$ and $c\alpha \in \mathcal{V}$ with $c \in \mathbb{Q}$, then $c \in \{\pm 1, \pm 1/2, \pm 2\}$.

1.2 Fundamental systems and Weyl chambers ([2], exposé 11,12,14)

Throughout this section, Φ is a root system of a free module X of rank ℓ, and W is the Weyl group of Φ.

Fix a linear order (compatible with the addition) in X, and denote by Φ_+ the set of all positive roots.

Definition: A positive root α is said to be simple if α cannot be expressed in the form $\beta + \gamma$ with $\beta, \gamma \in \Phi_+$.

The set of all simple roots of Φ is denoted Δ and is called a fundamental system of Φ.

Proposition 1.2.1: The fundamental system Δ consists of ℓ linearly independent roots $\alpha_1, \ldots, \alpha_\ell$. Every root $\alpha \in \Phi$ can be expressed uniquely in the form $\alpha = \pm \Sigma_{i=1}^{\ell} m_i \alpha_i$ with $m_i \in \mathbb{Z}$, $m_i \geq 0$.

We will sketch the proof of this proposition; to do so, we state two lemmas.[5)]

Lemma 1: Let α, $\beta \in \Phi$ be linearly independent, and put sign $\langle\alpha,\beta\rangle = \varepsilon$. Then $\beta - \varepsilon\alpha \in \Phi$. (Actually, it follows by an easy induction that $\beta - \varepsilon\alpha, \beta - 2\varepsilon\alpha, \ldots, \beta - c_{\alpha\beta}\alpha \in \Phi$.)

Lemma 2: Let V be a vector space with a metric $\langle\ ,\ \rangle$, $\{\alpha_i\}$ a basis for V, and $\{\xi_i\}$ a dual basis. If $\langle\alpha_i,\alpha_j\rangle \leq 0$ for all $i \neq j$, then $\langle\xi_i,\xi_j\rangle \geq 0$ for all $i \neq j$.

Outline of proof of Prop. 1.2.1

We define $\alpha_1, \ldots, \alpha_\ell$ as follows:

$$\alpha_1 = \min\{\Phi_+\},\ \alpha_i = \min(\Phi_+ - \{\alpha_1, \ldots, \alpha_{i-1}\}_{\mathbb{Q}}),\quad 2 \leq i \leq \ell.$$

It is clear from the definition that $\alpha_1, \ldots, \alpha_\ell$ are linearly independent, and $\alpha_1 < \alpha_2 < \cdots < \alpha_\ell$, so that $\langle\alpha_i,\alpha_j\rangle \leq 0$ if $i \neq j$ (Lemma 1). If we denote by $\{\xi_i\}$ the dual basis to $\{\alpha_i\}$ in X, then $<\xi_i,\xi_j> \geq 0$ for all $i \neq j$ (Lemma 2). To complete the proof, it suffices to show that if $\alpha \in \mathcal{V}_+$, then α can be written $\alpha = \Sigma_{i=1}^{\ell} m_i\alpha_i$, $m_i \geq 0$, $m_i \in \mathbb{Z}$ (this also implies the α_i are simple). Let $\alpha \in \mathcal{V}_+$. By induction on ℓ (since $\mathcal{V} \cap \{\alpha_1,\ldots,\alpha_k\}_{\mathbb{Q}}$ is a root system of the k-dimensional space $\{\alpha_1,\ldots,\alpha_k\}_{\mathbb{Q}}$), we may assume $\alpha \notin \{\alpha_1,\ldots,\alpha_{\ell-1}\}_{\mathbb{Q}}$; hence $\alpha \geq \alpha_\ell$. If, for some i, $\alpha - \alpha_i \in \mathcal{V}_+$, then by an induction on the linear order in X, we are done. Suppose $\alpha - \alpha_i \notin \mathcal{V}_+$ for all i. Then, since $\alpha \geq \alpha_\ell$, one has $\alpha - \alpha_i \notin \mathcal{V}$ for all i, so that $<\alpha_i,\alpha> \leq 0$ for all i (Lemma 1). But then it follows that

$$\alpha = \sum_i <\alpha,\alpha_i>\xi_i = \sum_{i,j} <\alpha,\alpha_i><\xi_i,\xi_j>\alpha_j \leq 0,$$

which is a contradiction.

<u>Remark</u>: Proposition 1.2.1 characterizes a fundamental system of $\mathcal{V}$; that is, if Δ is a subset of $\mathcal{V}$ satisfying the conditions of the proposition, then there exists a linear order on X such that Δ is the set of all simple roots (with respect to that order).

<u>Proposition 1.2.2</u> ([2],exposé 14)

1) Every root $\alpha \in \mathcal{V}$ can be written in the form $\alpha = w_{\alpha_{i_r}}\cdots w_{\alpha_{i_1}}\alpha_{i_0}$ for some $\alpha_{i_0} \in \Delta$, $\alpha_{i_1},\ldots,\alpha_{i_r} \in \Delta$.

2) W is generated by $\{w_{\alpha_i} \mid \alpha_i \in \Delta\}$.

Property 1) of this proposition (which can be proved by an argument similar to the one above) shows that the root system $\mathcal{V}$ is uniquely determined by the fundamental system Δ. Property 2) follows from 1) and the easily verified relation

$$w_{w_\alpha(\beta)} = w_\alpha w_\beta w_\alpha \quad \text{for all } \alpha, \beta \in \mathcal{V}.$$

Since W maps $\mathcal{V}$ onto itself, it is clear that for any $w \in W$, the set $w(\Delta)$ is another fundamental system of $\mathcal{V}$ (corresponding to a different linear order on X). Thus the group W acts on the set of all fundamental systems of $\mathcal{V}$ (corresponding to all possible linear orders on X). This action of W satisfies the following

<u>Lemma 3</u>: The only element of W which maps Δ onto itself is the identity.

Using this lemma, and the geometric notion of Weyl chamber (described below), one can prove

<u>Proposition 1.2.3</u>: W acts simply transitively on the set of all fundamental systems of $\mathcal{V}$.

The concept of Weyl chamber is introduced as follows. For each $\alpha \in \mathcal{V}$, define $H_\alpha = \{\chi \in X_{\mathbb{R}} \mid \langle \alpha, \chi \rangle = 0\}$; H_α is the hyperplane in $X_{\mathbb{R}}$ determined by α. The space $X_{\mathbb{R}} - \bigcup_{\alpha \in \mathcal{V}} H_\alpha$ is a finite union of disjoint connected components; such a connected component is called a Weyl chamber. It is easily shown that the set

$$\Lambda_\Delta = \{\chi \in X \mid \langle \alpha_i, \chi \rangle > 0,\ 1 \leq i \leq \ell\}$$

is a Weyl chamber, and in fact,

$$X_{\mathbb{R}} - \bigcup_{\alpha \in \mathcal{V}} H_\alpha = \bigcup_{\Delta:\text{f.s.}} \Lambda_\Delta$$

(the union on the right is taken over the set of all fundamental systems of $\mathcal{V}$). Since the metric $\langle\ ,\ \rangle$ on $X_{\mathbb{R}}$ is W-invariant, W leaves invariant the set of hyperplanes H_α, and W permutes the Weyl chambers as follows:

$$w(\Lambda_\Delta) = \Lambda_{w(\Delta)}.$$

By an easy topological argument, one sees that $\bigcup_{\Delta:\text{f.s.}} \Lambda_\Delta = \bigcup_{w \in W} w(\Lambda_\Delta)$, so that W acts transitively on the set of Weyl chambers.

Using the one-to-one correspondence between fundamental systems of $\surd$ and Weyl chambers (and Lemma 3), Proposition 1.2.3 follows.

<u>Definition</u>: A subset $\surd_1 \subset \surd$ is <u>closed</u> if $\surd_1 = \{\surd_1\}_{\mathbb{Z}} \cap \surd$. A subset $\surd_1 \subset \surd$ is $\mathbb{Q}$-<u>closed</u> if $\surd_1 = \{\surd_1\}_{\mathbb{Q}} \cap \surd$.

It is easily verified that if $\surd_1$ is closed, then $\surd_1$ is a root system in $X_1 = \{\surd_1\}_{\mathbb{Q}} \cap X$; $\surd_1$ is called a <u>subsystem</u> of $\surd$. If $\surd_1$ is $\mathbb{Q}$-closed, then there exists a fundamental system Δ of $\surd$ such that $\Delta_1 = \Delta \cap X_1$ is a fundamental system of $\surd_1$. To show how such a fundamental system Δ may be chosen, note that since X_1 is a submodule of X, then one can choose a basis $\xi_1,\ldots,\xi_k,\xi_{k+1},\ldots,\xi_\ell$ of $X_{\mathbb{Q}}$ so that $\xi_{k+1},\ldots,\xi_\ell$ is a basis of $X_{1\mathbb{Q}}$. The lexicographic order on X with respect to this basis of $X_{\mathbb{Q}}$, namely, $\chi > 0 \Leftrightarrow \chi = \sum_{i=j}^{\ell} \nu_i \xi_i$ and $\nu_j > 0$, satisfies the condition:

(7) if $\chi > 0$ and $\chi \notin X_1$, and $\chi' \equiv \chi \bmod(X_1)$, then $\chi' > 0$.

If Δ is the fundamental system of $\surd$ with respect to this order, and Δ_1 is the fundamental system of $\surd_1$ with respect to the order induced on X_1, then $\Delta_1 \subset \Delta \cap X_1$. (For, if $\alpha \in \Delta_1$ and α is not simple in $\surd$, then $\alpha = \alpha' + \alpha''$, $\alpha',\alpha'' \in \surd_+$, $\alpha',\alpha'' \notin X_1$, and $\alpha' = -\alpha''$ (mod X_1), which contradicts (7).) Clearly $\Delta \cap X_1 \subset \Delta_1$, so $\Delta_1 = \Delta \cap X_1$.

<u>Definition</u>: $\surd$ is <u>reducible</u> if $\surd = \surd_1 \cup \surd_2$ where $\surd_1, \surd_2$ are non-empty subsystems of $\surd$, and $\surd_1 \perp \surd_2$ ($\langle\alpha,\beta\rangle = 0$ for all $\alpha \in \surd_1$, $\beta \in \surd_2$). If $\surd$ is not reducible, it is called <u>irreducible</u>.

If $\surd$ is reducible, then one can decompose $X_{\mathbb{Q}}$ into orthogonal subspaces $X_{\mathbb{Q}} = \{\surd_1\}_{\mathbb{Q}} + \{\surd_2\}_{\mathbb{Q}}$, so that $\surd_i$ is $\mathbb{Q}$-closed, $i = 1,2$. Clearly every root system $\surd$ can be decomposed into a disjoint union of irreducible, mutually orthogonal subsystems: $\surd = \surd_1 \cup \ldots \cup \surd_s$, and this induces a decomposition of the fundamental system Δ of $\mathcal{V}$: $\Delta = \Delta_1 \cup \ldots \cup \Delta_s$, where Δ_i is a fundamental system of $\mathcal{V}_i$. We will say a fundamental system Δ of a root system $\mathcal{V}$ is irreducible if $\mathcal{V}$ is irreducible. Thus if $\Delta = \{\alpha_1,\ldots,\alpha_\ell\}$ is an irreducible fundamental system, Δ satisfies: i) $\alpha_1,\ldots,\alpha_\ell$ are linearly independent; ii) $\frac{2\langle\alpha_i,\alpha_j\rangle}{\langle\alpha_i,\alpha_i\rangle}$ is a non-positive integer if $i \neq j$; iii) Δ is not decomposable into two mutually orthogonal subsets.

The collection of finite sets of vectors $\Delta = \{\alpha_1,\ldots,\alpha_\ell\}$ in Euclidean space satisfying properties i), ii), iii) can be classified by use of the _Dynkin diagram of_ Δ. Given Δ, to each vector $\alpha_i \in \Delta$ associate a vertex, and connect the two vertices associated to α_i and α_j if and only if $\langle\alpha_i,\alpha_j\rangle \neq 0$. In our case, we have

$$c_{\alpha_i,\alpha_j} = \frac{2\langle\alpha_i,\alpha_j\rangle}{\langle\alpha_i,\alpha_i\rangle} \in \{0,-1,-2,-3\}$$

(since the Schwarz inequality implies $0 \leq \frac{2\langle\alpha_i,\alpha_j\rangle}{\langle\alpha_i,\alpha_i\rangle}\frac{2\langle\alpha_i,\alpha_j\rangle}{\langle\alpha_j,\alpha_j\rangle} \leq 4$). Thus we connect vertices corresponding to α_i and α_j with a single, double, or triple line according to whether $c_{\alpha_i,\alpha_j} = -1, -2, -3$ respectively. The arrows point from a longer to a shorter vector, when the lengths are different. The following is a complete list of the Dynkin diagrams of fundamental systems Δ satisfying the conditions i), ii), iii).[6)]

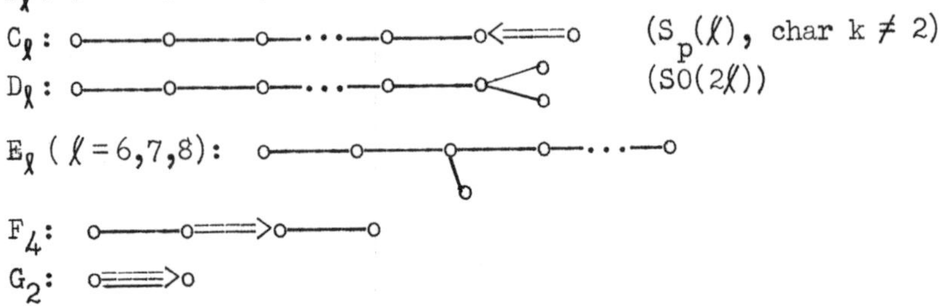

By the Chevalley's theorems in next section, it will be seen that this classification of Dynkin diagrams is equivalent to that of simple algebraic groups over algebraically closed fields.

1.3 The fundamental theorem of Chevalley ([2], exposés 17 ~ 24)

In this section, we summarize the main results which give the classification of semi-simple algebraic groups, due to Chevalley. These results are later assumed as the classification problem relative to a field of definition is considered (§2).

If G and G' are connected semi-simple algebraic groups and φ is an isogeny from G onto G', then φ induces a one-to-one correspondence between the set of closed connected subgroups of G and G',

$$H \rightarrow \varphi(H) = H', \quad [\varphi^{-1}(H)]^\circ \leftarrow H',$$

where $H \subset G$, $H' \subset G'$, and $[\varphi^{-1}(H)]^\circ$ is the connected component of the identity in $\varphi^{-1}(H)$. In particular, φ induces a one-to-one correspondence between the set of maximal tori of G and the set of maximal tori of G'. Let T be a maximal torus of G; $\varphi | T$ is an isogeny of T onto $T' = \varphi(T)$. If X, X' are the character modules of T, T' respectively, then the mapping ${}^t(\varphi|T)$ is an injection of X' into X, having finite co-kernel (see I, §2.4). We will abbreviate ${}^t(\varphi|T)$ as ${}^t\varphi$. Let $\mathfrak{r} \subset X$ and $\mathfrak{r}' \subset X'$ be root systems of G and G' (with respect to T and T', respectively), and for each $\alpha \in \mathfrak{r}$, $\alpha' \in \mathfrak{r}'$, let x_α and $x'_{\alpha'}$ be (fixed) isomorphisms of $\mathbf{G}_a$ onto the subgroups P_α and $P'_{\alpha'}$ of G and G' respectively, satisfying condition (1) (see §1.1). Since $tx_\alpha(\xi)t^{-1} = x_\alpha(\alpha(t)\xi)$ for all $t \in T$, $\xi \in \mathbf{G}_a$, we have, applying φ to both sides of the equation, $\varphi(t)(\varphi \circ x_\alpha)(\xi)\varphi(t)^{-1} = (\varphi \circ x_\alpha)(\alpha(t)\xi)$. Thus $\varphi(P_\alpha)$ is a one-dimensional unipotent subgroup of G' which is invariant under T', and so there is a unique root $\alpha' \in \mathcal{V}'$ such that $P'_{\alpha'} = \varphi(P_\alpha)$ ([2],18-03, lemma 1). The mapping φ induces an isogeny ψ of $\mathbb{G}_a$ onto itself, namely ψ is the mapping which makes the diagram below commute:

$$\begin{array}{ccc} \mathbb{G}_a & \xrightarrow{x_\alpha} & P_\alpha \\ \psi\downarrow & & \downarrow\varphi \\ \mathbb{G}_a & \xrightarrow{x'_{\alpha'}} & P'_{\alpha'} \end{array}$$

It can be shown that $\psi(\xi) = \lambda\xi^{q_\alpha}$ for all $\xi \in \mathbb{G}_a$, where $\lambda \in \Omega$, and q_α is a power of the characteristic exponent p[2)]. Using this diagram, and property (1), and denoting $t' = \varphi(t)$ for $t \in T$, we have:

$$\begin{aligned} x'_{\alpha'}(\alpha'(t')\lambda\xi^{q_\alpha}) &= t'(x'_{\alpha'}\circ\psi)(\xi)t'^{-1} = t'(\varphi\circ x_\alpha)(\xi)t'^{-1} \\ &= (\varphi\circ x_\alpha)(\alpha(t)\xi) \qquad = (x'_{\alpha'}\circ\psi)(\alpha(t)\xi) \\ &= x'_{\alpha'}(\lambda\alpha(t)^{q_\alpha}\,\xi^{q_\alpha}), \quad \text{for all } t \in T,\ \xi \in \mathbb{G}_a. \end{aligned}$$

Since $\alpha'(t') = [{}^t\varphi(\alpha')](t)$, the previous equation implies $[{}^t\varphi(\alpha')](t) = \alpha(t)^{q_\alpha}$ for all $t \in T$, or, using additive notation,

(8) $$ {}^t\varphi(\alpha') = q_\alpha\,\alpha\,. $$

<u>Definition</u>: (G, T, X, $\mathcal{V}$ and G', T', X', $\mathcal{V}'$ as above).

An injective homomorphism $\rho: X'_{\mathbb{Q}} \to X_{\mathbb{Q}}$ is called <u>special</u> if

i) $\rho(X') \subset X$;

ii) there exists a bijection f of $\mathcal{V}$ onto $\mathcal{V}'$ such that $(\rho\circ f)(\alpha) = q_\alpha\alpha$ for every root $\alpha \in \mathcal{V}$, where q_α is a power of p, (p is the characteristic exponent.)

What we have proved above is that the injection ${}^t\varphi$ determined by the isogeny φ of G onto G' is special. The converse of this result is difficult to prove, and is the fundamental theorem of Chevalley. Before stating this, we note that as a consequence of (8), we have

$$\deg \varphi = [X : {}^t\varphi(X')] \prod_{\alpha \in \mathcal{V}} q_\alpha . \tag{9}$$

In particular, φ is an isomorphism of G onto G' if and only if ${}^t\varphi$ is surjective and $q_\alpha = 1$, for all $\alpha \in \mathcal{V}$. (Property (9) is a consequence of the "Bruhat Decomposition" of G, G' which implies G is birationally equivalent to the direct product $T \times \coprod_{\alpha \in \mathcal{V}} P_\alpha$ ($\coprod$ = direct product), and G' birationally equivalent to $T' \times \coprod_{\alpha' \in \mathcal{V}} P'_{\alpha'}$. This implies $\Omega(G)$ (resp., $\Omega(G')$) is the product of the function fields $\Omega(T)$, $\Omega(P_\alpha)$, $\alpha \in \mathcal{V}$ (resp., $\Omega(T')$, $\Omega(P'_{\alpha'})$, $\alpha' \in \mathcal{V}'$) so that $\deg \varphi$ is the product of $\deg(\varphi \mid T)$, $\deg(\varphi \mid P_\alpha)$, $\alpha \in \mathcal{V}$. Since $\deg(\varphi \mid T) = [X : {}^t\varphi(X')]$, and $\deg(\varphi \mid P_\alpha) = q_\alpha$, $\alpha \in \mathcal{V}$, (9) follows.)

<u>Theorem 1.3.1</u> (Fundamental Theorem of Chevalley) ([2], exposé 23):

Let G, G' be connected semi-simple algebraic groups, T, T' maximal tori of G, G' respectively, and X, X' the character modules of T, T'. If there exists a special injective homomorphism $\rho : X' \to X$, then there exists an isogeny $\varphi : G \to G'$ such that ${}^t\varphi = \rho$. Moreover, φ is unique up to inner automorphism defined by $t \in T$.

<u>Theorem 1.3.2</u> (Existence Theorem of Chevalley):

Let k_o be any prime field. If X is a free module of finite rank, and $\mathcal{V}$ a root system in X, then there exists a connected semi-simple algebraic group G, defined over k_o, having $(X, \mathcal{V})$ as its root system (with respect to some maximal torus T of G). Moreover, T can be taken to be k_o-trivial.

A group G satisfying theorem 1.3.2 is called a "Chevalley group"; by Theorem 1.3.1, such a group is uniquely determined by $(X, \mathcal{V})$, up to an isomorphism defined over k_o. For the proof of theorem 1.3.2, see

Chevalley, "Sur Certains Groupes Simples," Tohoku Math. Jour., 1955, pp. 14-66 (the proof is an actual construction of G).[7] We will denote the Chevalley group determined by $(X, \surd)$ as $G(X, \surd)$.[7a] Later we will prove that if G is a semi-simple connected group defined over k (k perfect) and G has a k-trivial maximal torus T with a root system $(X, \surd)$, then G is in fact unique up to k-isomorphism.

Before examining consequences of these two fundamental theorems, we list some examples.

<u>Example 1</u>: Let $(X, \surd)$ be a root system. The module $X_o = \{\surd\}_{\mathbb{Z}}$ is called the root module of $\surd$, and the module $X^o = \{\surd^*\}^{\wedge}_{\mathbb{Z}}$ is called the weight module of $\surd$. By property (6) (§1.1) we have natural injections $X_o \subset X \subset X^o$. Both injections are special, where the corresponding bijection f of $\surd$ onto $\surd$ is the identity, and $q_\alpha = 1$, all $\alpha \in \surd$. Thus by Theorems 1.3.1, 1.3.2, there exist isogenies

$$G(X^o, \surd) \to G(X, \surd) \to G(X_o, \surd).$$

An isogeny φ which satisfies $q_\alpha = 1$, all $\alpha \in \surd$ (q_α as in (8)) is called a <u>standard</u> isogeny; thus the above isogenies are standard. The group $G(X^o, \surd)$ is called the <u>simply connected group</u> of type $\surd$; $G(X^o, \surd)$ has the largest possible character module (namely, X^o) for the given root system $\surd$. The group $G(X_o, \surd)$ is called the <u>adjoint group</u> of type $\surd$; $G(X_o, \surd)$ has the smallest possible character module for the given root system $\surd$. The simply connected group $G(X^o, \surd)$ has the following "universal" property: if $\tilde{G}$ is a semi-simple connected algebraic group and φ a standard isogeny of $\tilde{G}$ onto $G(X, \surd)$, then there exists an isogeny ρ which makes the following diagram commute:

$$\begin{array}{ccc} \tilde{G} & \xrightarrow{\varphi} & G(X, \surd). \\ \rho \uparrow & \nearrow & \\ G(X^o, \surd) & & \end{array}$$

In particular, when the universal domain is $\mathbb{C}$, $G(X^o, \mathcal{V})$ is a universal covering group of $G(X, \mathcal{V})$ in the sense of topological groups.

Example 2: Let G be a semi-simple connected algebraic group, defined over $k = \mathbb{F}_q$, and φ_q the Frobenius endomorphism of G onto itself (see I, §2.2). Let T be a maximal torus of G. Referring to the Example in I, §2.4, we see that ${}^t\varphi_q(\alpha') = q\alpha$ (here $\alpha' = \alpha^{(q)}$), so that $q_\alpha = q$ for all $\alpha \in \mathcal{V}$. From (9) it follows that $\deg \varphi_q = q^{\dim G}$.

Example 3: In three classic cases, "singular isogenies" occur; a singular isogeny satisfies the condition that $\min(q_\alpha) = 1$, but $\max(q_\alpha) \neq 1$. If the prime field has characteristic 2, then there are singular isogenies $B_\ell \leftrightarrows C_\ell$, $F_4 \leftrightarrows F_4$; in characteristic 3, there is a singular isogeny $G_2 \leftrightarrows G_2$. We indicate only the case $B_\ell \leftrightarrows C_\ell$ (char. 2). The group B_ℓ is $SO(2\ell + 1)$, the special orthogonal group with respect to the form

$$S = \left(\begin{array}{cc|c} 0 & 1_\ell & \\ 1_\ell & 0 & 0 \\ \hline & 0 & 1 \end{array}\right).$$

A maximal torus T in B_ℓ is the group of all diagonal matrices

$$t = \left(\begin{array}{c|c|c} \begin{matrix}\xi_1 & & \\ & \ddots & \\ & & \xi_\ell\end{matrix} & & \\ \hline 0 & \begin{matrix}\xi_1^{-1} & & \\ & \ddots & \\ & & \xi_\ell^{-1}\end{matrix} & 0 \\ \hline & 0 & 1 \end{array}\right).$$

If χ_i is the character defined by $\chi_i(t) = \xi_i$, $1 \leq i \leq \ell$, then the root system $\mathcal{V}$ of B_ℓ with respect to T is $\mathcal{V} = \{\chi_i \pm \chi_j \ (i \neq j), \pm \chi_i\}$. From the classification of simple groups ([2], exposé 19), it is known that the set $\mathcal{V}' = \{\chi_i \pm \chi_j \ (i \neq j), \pm 2\chi_i\}$ is a root system of an algebraic group of type C_ℓ. If X(T) is ordered lexicographically with respect to

$\lambda_1,\dots,\lambda_\ell$, then the fundamental systems Δ, Δ' of $\surd$, $\surd'$ respectively are $\Delta = \{\lambda_1-\lambda_2,\dots,\lambda_{\ell-1}-\lambda_\ell,\ \lambda_\ell\}$, $\Delta' = \{\lambda_1-\lambda_2,\dots,\lambda_{\ell-1}-\lambda_\ell,\ 2\lambda_\ell\}$. Let X_o and Y_o be the root modules of $\surd$ and $\surd'$ respectively. Then $X_o = \{\lambda_1,\dots,\lambda_\ell\}_{\mathbb{Z}}$ and $Y_o = \{\Sigma\, m_i\, \lambda_i \in X_o \mid \Sigma\, m_i \equiv 0 \pmod 2\}$, so that $[X_o : Y_o] = 2$. Now if we identify $\hat{X}_o$ with X_o by means of the standard inner product, we have $\surd^* = \surd'$, $\surd'^* = \surd$. Hence the weight modules X^o, Y^o of $\surd$ and $\surd'$ are by definition the duals $\hat{Y}_o$, $\hat{X}_o$ of Y_o, X_o respectively. Thus one has $Y^o = \hat{X}_o = X_o$, $X^o = \hat{Y}_o = \{\lambda_1,\dots,\lambda_\ell,\ \frac{1}{2}\Sigma_{i=1}^{\ell}\ \lambda_i\}_{\mathbb{Z}}$, and the standard isogenies $G(X^o,\surd) \to G(X_o,\surd)$, $G(Y^o,\surd') \to G(Y_o,\surd')$ are of degree 2. When the prime field has characteristic 2, it can be shown that there are singular isogenies φ, ψ, η which make the diagram below commute:

$$\begin{array}{ccc} G(X^o,\surd) & \longrightarrow & G(X_o,\surd) \\ \uparrow\varphi & \swarrow\psi & \uparrow\eta \\ G(Y^o,\surd') & \longrightarrow & G(Y_o,\surd'). \end{array}$$

The isogeny ψ is defined by the identification $Y^o = X_o$, which induces the natural correspondence between the fundamental systems

Δ

Δ'

Thus for ψ, the corresponding q_α for $\alpha \in \Delta$ are $q_{\alpha_i} = 1$, $1 \le i \le \ell-1$, $q_{\alpha_\ell} = 2$, and $[X_o : {}^t\psi(Y^o)] = 1$. The isogenies φ and η are determined by the commutativity of the diagram; in particular, $[Y^o : {}^t\varphi(X^o)] = 2^{\ell-1}$, $[Y_o : {}^t\eta(X_o)] = 2^{\ell-1}$, and the q_α corresponding both to φ and η for $\alpha \in \Delta$ are: $q_{\alpha_i} = 2$, $1 \le i \le \ell-1$, $q_{\alpha_\ell} = 1$.

The singular isogenies $F_4 \rightleftarrows F_4$ (char. 2), and $G_2 \rightleftarrows G_2$ (char. 3) are defined in a similar manner, and correspond to the mappings of the Dynkin diagrams:

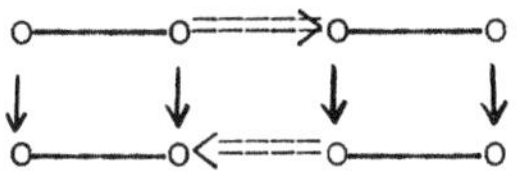 (F_4) 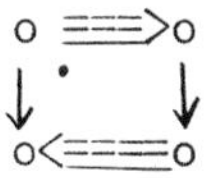 (G_2)

Recently, new finite simple groups have been discovered through the study of these singular isogenies. According to Tits ("Algebraic and Abstract Simple Groups," Annals of Math., Vol. 80 (1964)) if G is a k-simple algebraic group (G is defined over k, but has no proper normal subgroups defined over k), and G is not k-compact, then $D(G_k)/C(G_k)$ is a simple group (except in a few cases), where $D(G_k)$ is the derived group of G_k, and $C(G_k)$ the center of G_k (k an arbitrary field). In particular, if k is a finite field, one obtains in this manner a finite simple group of Lie type. For the case $G = B_2$, and $k = \mathbb{F}_q$, char $k = 2$, the singular isogeny $B_2 \rightleftarrows C_2$ just constructed induces a singular endomorphism of $G = B_2$, whose restriction to G_k is an automorphism (actually, an involution of G_k). The subgroup of elements of G_k left fixed by this involution is a simple group, called a Suzuki group. (See M. Suzuki, "A New Type of Simple Groups of Finite Order," Proc. Nat. Acad. U.S.A., Vol. 46 (1960), also T. Ono, "An Identification of Suzuki Groups with Groups of Generalized Lie Type," Annals of Math., Vol. 75 (1962).) In a similar manner, from the singular isogenies $F_4 \rightleftarrows F_4$ (char. 2), $G_2 \rightleftarrows G_2$ (char. 3), new finite groups were discovered by Ree. (See R. Ree, "A family of simple groups associated with the simple Lie algebra (G_2)" (resp. "...(F_4)"), Bulletin A.M.S., Vol. 60 (1960), (resp. Vol. 61 (1961).)

1.4 Some consequences of the fundamental theorem of Chevalley

We wish first to determine the structure of Aut(G), where G is a Chevalley group. Denote by Inn(G) the group of inner automorphisms of G; clearly under the correspondence $g_1 \longleftrightarrow [I_{g_1}: g \to g_1 g g_1^{-1}]$, Inn(G) can be identified with G/Z (Z = center of G). Also, $\mathrm{Inn}(G)$ is a normal subgroup of $\mathrm{Aut}(G)$.

<u>Proposition 1.4.1</u>: Let G be a Chevalley group. There exists a finite subgroup $\Theta \subset \mathrm{Aut}(G)$ such that $\mathrm{Aut}(G) = \Theta\cdot\mathrm{Inn}(G)$ (semi-direct product).[8]

We outline the proof; see [2], exposé 17. Let T be a maximal torus of G, $X = X(T)$, $\surd$ a root system of G with respect to T, and Δ a fundamental system of $\surd$. Define $\Theta' = \{\theta \in \mathrm{Aut}(G) \mid \theta(T) = T,\ {}^t\theta(\Delta) = \Delta\}$. Clearly Θ' is a subgroup of $\mathrm{Aut}(G)$, and $\Theta' \supset T/Z$. Let $\varphi \in \mathrm{Aut}(G)$. Since $\varphi(T)$ is a maximal torus of G, $\varphi(T) = I_{g_1}(T)$ for some $g_1 \in G$ (I, §4.3, Remark 1). Let $\varphi_1 = I_{g_1}^{-1}\circ\varphi$; then $\varphi_1(T) = T$, so ${}^t\varphi_1 \in \mathrm{Aut}(X,\surd)$ (in general, if M is any subset of X, we denote $\mathrm{Aut}(X,M) = \{\psi \in \mathrm{Aut}(X) \mid \psi(M) = M\}$). This implies ${}^t\varphi_1(\Delta) = \Delta'$ is a fundamental system of $\surd$, so there exists a unique $w \in W$ satisfying $w(\Delta) = \Delta'$ (Prop. 1.2.3). Since $w = {}^t(I_s|T)^{-1}$ for some $s \in N(T)$, it follows that ${}^tI_s\circ{}^t\varphi_1(\Delta) = \Delta$, so $\varphi_2 = \varphi_1\circ I_s \in \Theta'$, and thus $\varphi \in \Theta'\cdot\mathrm{Inn}(G)$. Now $\Theta' \cap \mathrm{Inn}(G) = T/Z$, for if $\theta \in \Theta'$ and $\theta = I_s$, then $s \in N(T)$, and ${}^t\theta = w_s^{-1}$, but $w_s^{-1}(\Delta) = \Delta$ implies $w_s^{-1} = 1$ (Lemma 3, §1.2), so $s \in T$. Clearly T/Z is a normal subgroup of Θ', and since ${}^t\theta$ determines θ modulo T/Z (Thm. 1.3.1) and Δ is finite, it follows that $\Theta'/T/Z$ is a finite group.

The subgroup $\Theta \subset \mathrm{Aut}(G)$ we seek is isomorphic to the quotient group $\Theta'/T/Z$. To prove its existence, we must show this quotient group splits. We will do this in §2.4.1. The group $\Theta'/T/Z$ can be identified with $\mathrm{Aut}(X,\Delta)$, which is studied case-by-case (one examines the Dynkin diagrams, §1.2). The only cases where $\mathrm{Aut}(X,\Delta)$ is not just the identity are type A_ℓ ($\mathrm{Aut}(X,\Delta)$ has order 2, the non-trivial automorphism being o—o—...—o—o, with arrows exchanging the nodes symmetrically), type D_ℓ (if $\ell > 4$, $\mathrm{Aut}(X,\Delta)$ has order 2, the non-trivial automorphism being , and if $\ell = 4$, $Aut(X,\Delta) = \mathfrak{S}_3$, the symmetric group on three letters, corresponding to all permutations of the 'outer' vertices of the diagram 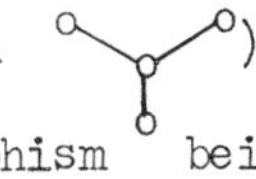), and type E_6 ($Aut(X,\Delta)$ has order 2, the non-trivial automorphism being

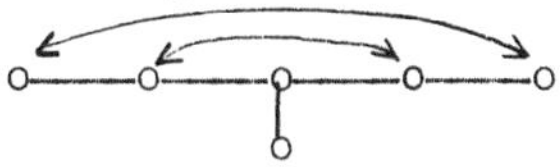).

Remark 1: In the outline of the proof of Prop. 1.4.1, we have displayed the following lattice:

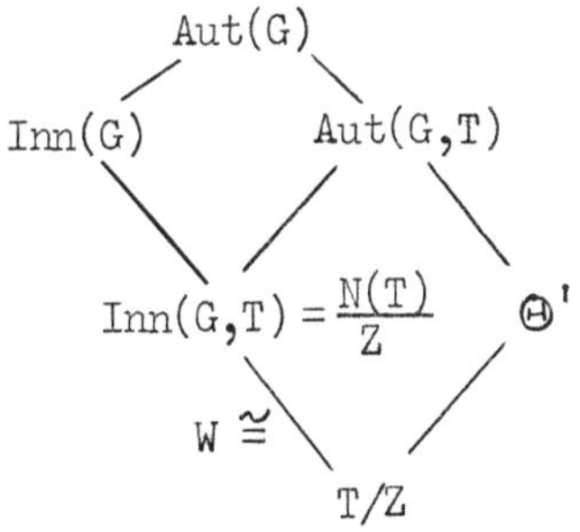

Remark 2: The group $Aut(X,\surd)$ is called the Cartan group of G. A special case of Thm. 1.3.1 says that any element in $Aut(X,\surd)$ is of the form ${}^t\varphi$ for some $\varphi \in Aut(G,T)$. Thus Prop. 1.4.1 (and the lattice above) shows that the Cartan group splits into the semi-direct product of W with $Aut(X,\Delta)$.

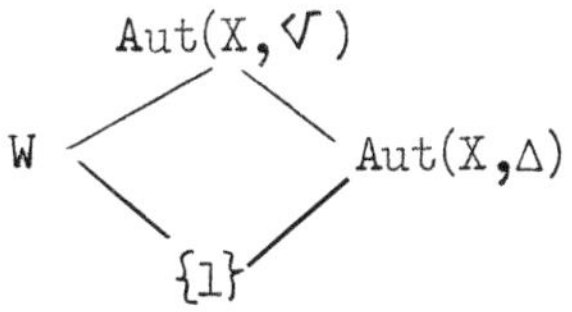

Remark 3: Another application of Theorem 1.3.1 occurs in reducing the classification problem to irreducible root systems. Let $\surd$ be a root system in a module X of finite rank, and let $\surd = \surd_1 \cup \ldots \cup \surd_s$ be the decomposition of $\surd$ into mutually orthogonal irreducible subsystems. Define $X_i = X \cap \{\surd_i\}_{\mathbb{Q}}$. The natural injection $(\oplus X_i, \cup \surd_i) \to (X,\surd)$

is special, so by Thm. 1.3.1, there exists a corresponding isogeny $G(X,\Psi) \to \prod_{i=1}^{s} G(X_i,\Psi_i)$ (direct product). Now in general, if $\Psi_1 \subset \Psi$ is a closed subsystem of Ψ, and $G(\Psi_1)$ denotes the closed subgroup of $G(X,\Psi)$ generated by $\{P_\alpha \mid \alpha \in \Psi_1\}$, then $G(\Psi_1)$ is a connected semi-simple algebraic subgroup containing the maximal torus $T(\Psi_1)$ generated by $\{\alpha^*(\mathbb{G}_m) \mid \alpha \in \Psi_1\}$ ($\alpha^* \in \hat{X} = \mathrm{Hom}(X,\mathbb{Z}) = \mathrm{Hom}(\mathbb{G}_m,T)$; this identification is a special case of the discussion after Lemma 2.4.3, Chapter I). The restriction of the isogeny $G(X,\Psi)\to\prod_{i=1}^{s} G(X_i,\Psi_i)$ to the subgroup $G(\Psi_i)$ gives an isogeny $G(\Psi_i)\to G(X_i,\Psi_i)$ (see [2], exposé 17). Note that when $G(X,\Psi)$ is a simple connected or adjoint group, the above isogeny $G(X,\Psi)\to\prod_{i=1}^{s} G(X_i,\Psi_i)$ actually becomes an isomorphism (see Example 1, §1.3).

For the remainder of this section, we wish to obtain results concerning the field of definition. The first proposition modifies Theorem 1.3.1.

Proposition 1.4.2[9]: Let $G(X,\Psi)$, $G(X',\Psi')$ be Chevalley groups defined over the prime field k_o, containing maximal tori T, T' respectively, which are k_o-trivial. If there exists a special injective homomorphism $\rho: X' \to X$, then there exists an isogeny φ/k_o: $G(X,\Psi) \to G(X',\Psi')$ such that ${}^t\varphi = \rho$.

Proof: By Theorem 1.3.1 (which is true over any algebraically closed field) there exists an isogeny $\varphi/\bar{k}_o$ of $G(X,\Psi)$ onto $G(X',\Psi')$ such that ${}^t\varphi = \rho$. If $\sigma \in \Gamma = \mathrm{Gal}(\bar{k}_o/k_o)$, then σ acts trivially on X and X' (Prop. 2.4.6), so that ${}^t(\varphi^\sigma) = {}^t\varphi = \rho$. By the uniqueness of φ (Theorem 1.3.1), it follows that $\varphi^\sigma = I_{t_\sigma}\circ\varphi$ for some $t_\sigma \in T'$. We may assume $t_\sigma \in T'_{\bar{k}_o}$; let $\bar{t}_\sigma$ denote the coset in $(T'/Z')_{\bar{k}_o}$ determined by t_σ. Identifying $(T'/Z')_{\bar{k}_o} = (\bar{k}_o{}^*)^\ell$, it is easily verified that $\bar{t}_\sigma$ is a one-cocycle of Γ

with values in $(\bar{k}_o*)^\ell$, so that by Hilbert's Theorem 90, $\bar{t}_\sigma$ is trivial. Hence there exists s ε T' such that $\bar{t}_\sigma = \bar{s}^\sigma \bar{s}^{-1}$; this implies $I_{t_\sigma} = I_s^\sigma \circ I_s^{-1}$. Thus if we replace φ by $I_s^{-1} \circ \varphi$, we see that $\varphi^\sigma = \varphi$ for all $\sigma \in \Gamma$, so φ is defined over k_o.

Corollary: The isogeny $G(X,\mathcal{V}) \to \prod G(X_i,\mathcal{V}_i)$ (Remark 3) can be taken to be defined over k_o. (The same is true for the isogenies discussed in 1.3.)

Next, we wish to consider the following problem: Let k be perfect; k_o its prime field. Given Chevalley groups and an isogeny $\varphi: G(X,\mathcal{V}) \to G(X',\mathcal{V}')$, all defined over k_o, and given (G,f) (a connected semi-simple algebraic group which is) a k-form of $G(X,\mathcal{V})$, what are necessary and sufficient conditions for the existence of a k-form (G',f') and an isogeny φ'/k which make the following diagram commute:

$$\begin{array}{ccc} G/k & \xrightarrow{f/\bar{k}} & G(X,\mathcal{V}) \\ \downarrow \varphi'/k & & \downarrow \varphi/k_o \\ G'/k & \xrightarrow{f'/\bar{k}} & G(X',\mathcal{V}'). \end{array}$$

To find these conditions, we may assume, without loss of generality, that T is a maximal torus of G defined over k, and f(T) is a k_o-trivial torus of $G(X,\mathcal{V})$. Now assume (G',f') and φ' exist, and denote (φ_σ) and (φ'_σ) the systems corresponding to the k-forms (G,f) and (G',f') respectively (see I, §3.1; recall that $\varphi_\sigma = f^\sigma \circ f^{-1}$, $\sigma \in \Gamma = \mathrm{Gal}(\bar{k}/k)$). We then have $\varphi \circ f = f' \circ \varphi'$, which implies

$$\varphi \circ f^\sigma = f'^\sigma \circ \varphi' \qquad \text{for all } \sigma \in \Gamma, \text{ hence}$$

$$(\star) \qquad \varphi \circ \varphi_\sigma = \varphi'_\sigma \circ \varphi \qquad \text{for all } \sigma \in \Gamma.$$

Since T and f(T) are defined over k, we have $\varphi_\sigma \in \mathrm{Aut}(G,f(T))$, ${}^t\varphi_\sigma \in \mathrm{Aut}(X,\mathcal{V})$ for all σ ε Γ. By duality (★) becomes ${}^t\varphi_\sigma \circ {}^t\varphi = {}^t\varphi \circ {}^t\varphi'_\sigma$, and thus

(10) ${}^t\varphi(X')$ is invariant under the automorphisms $\{{}^t\varphi_\sigma\}_{\sigma\in\Gamma}$.

We have seen (§1.3) that φ induces a one-to-one correspondence between $\sqrt{\Gamma}$ and $\sqrt{\Gamma}'$ (given by $\alpha \longleftrightarrow \alpha'$ where $\varphi(P_\alpha) = P'_{\alpha'}$), and that there exist numbers q_α satisfying ${}^t\varphi(\alpha') = q_\alpha\alpha$ for $\alpha \in \sqrt{\Gamma}$, $\alpha' \in \sqrt{\Gamma}'$ with $\alpha \longleftrightarrow \alpha'$. Since φ_σ and φ'_σ are automorphisms of G and G' respectively, ${}^t\varphi_\sigma$ and ${}^t\varphi'_\sigma$ are automorphisms of $(X,\sqrt{\Gamma})$ and $(X',\sqrt{\Gamma}')$ (i.e., the numbers q_α of equation (8) corresponding to these automorphisms are all equal to 1). Now the dual of (★) implies

$$({}^t\varphi \circ {}^t\varphi'_\sigma)(\alpha') = ({}^t\varphi_\sigma \circ {}^t\varphi)(\alpha') = {}^t\varphi_\sigma(q_\alpha\alpha) = q_\alpha\,{}^t\varphi_\sigma(\alpha).$$

Thus we have ${}^t\varphi_\sigma(\alpha) \longleftrightarrow {}^t\varphi'_\sigma(\alpha')$, and

(11) $q_{{}^t\varphi_\sigma(\alpha)} = q_\alpha$ for all $\alpha \in \sqrt{\Gamma}$, $\sigma \in \Gamma$, where q_α is as in equation (8).

<u>Proposition 1.4.3</u>: Conditions (10) and (11) are necessary and sufficient for the existence of (G',f') and φ' as described in the <u>problem</u> (p.61).

<u>Proof</u>: We only need to show (10) and (11) are sufficient. Since ${}^t\varphi$ is an injection of X' into X, condition (10) implies there exists a system (ρ_σ) of automorphisms of X' which satisfy

(★★) $${}^t\varphi_\sigma \circ {}^t\varphi = {}^t\varphi \circ \rho_\sigma \quad \text{for all } \sigma \in \Gamma.$$

Condition (11) implies that if $\alpha' \in \sqrt{\Gamma}'$, then

$${}^t\varphi(\rho_\sigma(\alpha')) = ({}^t\varphi_\sigma \circ {}^t\varphi)(\alpha') = {}^t\varphi_\sigma(q_\alpha\alpha) = q_{{}^t\varphi_\sigma(\alpha)}\,{}^t\varphi_\sigma(\alpha)$$

which shows that $\rho_\sigma(\alpha') \in \sqrt{\Gamma}'$, and ${}^t\varphi_\sigma(\alpha) \longleftrightarrow \rho_\sigma(\alpha')$. Since ρ_σ is an automorphism of $(X',\sqrt{\Gamma}')$, there exists a system (φ'_σ), $\varphi'_\sigma \in \mathrm{Aut}(G(X',\sqrt{\Gamma}'))$ satisfying ${}^t\varphi'_\sigma = \rho_\sigma$ (Theorem 1.3.1). Since φ'_σ is unique up to inner automorphism, (★★) implies (after adjusting φ'_σ by an inner automorphism if necessary) $\varphi \circ \varphi_\sigma = \varphi'_\sigma \circ \varphi$, for all $\sigma \in \Gamma$. Since (φ_σ) is a one-cocycle, this equation shows that (φ'_σ) is a one-cocycle of Γ with values in $\mathrm{Aut}(G(X', \nabla'))$, hence there exists a corresponding k-form (G', f') with $\varphi'_\sigma = f'^\sigma \circ f'^{-1}$ (Prop. 3.1.1). If we define $\varphi' = f'^{-1} \circ \varphi \circ f$, then φ' is defined over k, and the diagram commutes.

Remark 4: If the original problem is changed by reversing the direction of the vertical arrows in the diagram (i.e., φ/k_o: $G(X', \nabla')/k_o \to G(X, \nabla)/k_o$), then it can also be shown that conditions (10) and (11) are necessary and sufficient for the existence of a group G'/k which "covers" G.

Remark 5: We apply Prop. 1.4.3 and Remark 4 to the following diagram:

$$\begin{array}{ccccc} G(X^o, \nabla) & \longrightarrow & G(X, \nabla) & \longrightarrow & G(X_o, \nabla) \\ & & \uparrow{\scriptstyle f/\bar{k}} & & \\ & & G/k & & \end{array}$$

where the horizontal arrows are the standard isogenies described in Example 1, §1.3 (defined over k_o by Prop. 1.4.2) and (G, f) is a k-form of $G(X, \nabla)$. Since $X_o = \{\nabla\}_{\mathbb{Z}} \subset X \subset \{\nabla^*\}_{\mathbb{Z}}^{\wedge} = X^o$, it is clear that X_o and X are left invariant (in X and X^o respectively) by the operation of Γ, so that condition (10) is satisfied. Condition (11) is trivially satisfied, since the horizontal isogenies are standard. Thus there exist semi-simple connected groups G^o/k, G_o/k (called the simply connected covering group of G and the adjoint group of G, respectively), and k-isogenies φ^o, φ_o such that the following diagram commutes:

$$\begin{array}{ccccc} G(X^o, \nabla) & \longrightarrow & G(X, \nabla) & \longrightarrow & G(X_o, \nabla) \\ \uparrow{\scriptstyle f^o/\bar{k}} & & \uparrow{\scriptstyle f/\bar{k}} & & \uparrow{\scriptstyle f_o/\bar{k}} \\ G^o/k & \xrightarrow{\varphi^o} & G/k & \xrightarrow{\varphi_o} & G_o/k \, . \end{array}$$

We close this section with an outline of how the classification problem of connected semi-simple algebraic groups over a given perfect ground field k is reduced to the problem of classification of absolutely simple algebraic groups defined over K, where K/k is finite ("absolutely simple" means G has an irreducible root system).

Let G/k be a connected semi-simple algebraic group. From Theorems 1.3.1 and 1.3.2, it follows that G is $\bar{k}$-isomorphic to a Chevalley group $G(X, \check{V})$, where $(X, \check{V})$ is the root system of G with respect to a maximal torus T. By Remark 5, we may reduce to the case where G is simply connected. Hence we may suppose G/k is $\bar{k}$-isomorphic to the direct product πG_i, the G_i connected simple algebraic groups defined over $\bar{k}$ (Remark 3). Define $\Gamma_1 = \{\sigma \in \Gamma \mid G_1^{\sigma} = G_1\}$. Then Γ_1 determines an extension K_1 of k (K_1 = fixed field of Γ_1), and since G_1 is defined over some finite extension of k, it follows that K_1/k is finite. Let $d = [K_1:k]$, and $(\sigma_1, \ldots, \sigma_d)$ be a complete set of coset representatives of $\Gamma_1 \backslash \Gamma$ with σ_1 = identity. The set $\{G_1^{\sigma_1}, \ldots, G_1^{\sigma_d}\}$ is a complete set of conjugates of G_1 by Γ, and each of these groups is a direct factor of G as well, since any $\sigma \in \Gamma$ permutes the direct factors of G. Thus G is $\bar{k}$-isomorphic to the direct product $\prod_{i=1}^{d} G_1^{\sigma_i} \times G'$, and $\prod_{i=1}^{d} G_1^{\sigma_i}$ is Γ-invariant, so is defined over k. Actually we can see easily that G is k-isomorphic to $R_{K_1/k}(G_1) \times G'$. Repeating the above argument with G' replacing G, we see that G is k-isomorphic to the direct product $R_{K_1/k}(G_1) \times R_{K_1'/k}(G_1') \times \ldots$, with each direct factor defined over k. In particular, we see that G is k-simple if and only if $G \cong R_{K_1/k}(G_1)$ where G_1/K_1 and G_1 is absolutely simple. Thus the classification of G reduces first to the classification of k-simple groups which in turn reduces to the classification of absolutely simple groups defined over K, where K/k is finite.

§2. The structure of semi-simple algebraic groups defined over k (k perfect) ([8], [9])

Throughout this section, G will denote a connected semi-simple algebraic group defined over k, where k is a perfect field.

2.1 Restricted roots

As we have seen in the last section, there exists a Chevalley group $\underline{G}/k$ and a K-isomorphism f of G onto $\underline{G}$, where K/k is a finite Galois extension of k (i.e., (G,f) is a K/k-form of $\underline{G}$). Let T/k be a maximal torus of G (Remark 2, I, §4.2), $\underline{T}$ a k-trivial maximal torus of $\underline{G}$; we may assume that $f(T) = \underline{T}$. Let X and $\underline{X}$ be the character modules of T and $\underline{T}$ respectively, and $\sqrt{\ }$ and $\underline{\sqrt{\ }}$ root systems in X and $\underline{X}$ with respect to T and $\underline{T}$. Finally, denote $\Gamma = \mathrm{Gal}(K/k)$.

We have seen (Prop. 2.4.6) that each $\chi \in X$ is defined over K, hence X is a Γ-module, and that $\underline{X}$ is a trivial Γ-module. In addition, from the definition of root (§1.1), it is easily seen that $\sqrt{\ }$ is Γ-invariant; in fact, if $\alpha \in \sqrt{\ }$, then the one-dimensional unipotent subgroup of G corresponding to α^σ for $\sigma \in \Gamma$ is just $P_\alpha^{\ \sigma}$, the conjugate of P_α by σ. If, as usual, we denote $\varphi_\sigma = f^\sigma \circ f^{-1} \in \mathrm{Aut}_K(\underline{G},\underline{T})$, then ${}^t\varphi_\sigma \in \mathrm{Aut}(\underline{X},\underline{\sqrt{\ }})$, and if ${}^tf(\underline{\chi}) = \chi$, for $\underline{\chi} \in \underline{X}$, $\chi \in X$, then ${}^tf \circ {}^t\varphi_\sigma(\underline{\chi}) = \chi^\sigma$. Thus the automorphisms ${}^t\varphi_\sigma$ transpose the action of Γ on X to an action of Γ on $\underline{X}$. Often we will want to identify X and $\underline{X}$ (via the isomorphism ${}^tf$); when X is regarded as the character module of T, the action of Γ is non-trivial, and we identify $\chi^\sigma = {}^t\varphi_\sigma(\underline{\chi})$ (where $\chi = \underline{\chi}$ under the identification), and when X is regarded as the character module of $\underline{T}$, Γ acts trivially. Define

$$X_0 = \{\chi \in X \mid \sum_{\sigma \in \Gamma} \chi^\sigma = 0\} ,$$

$$X^\Gamma = \{\chi \in X \mid \chi^\sigma = \chi \quad \text{for all } \sigma \in \Gamma\}.$$

We have seen that X_0 and X^Γ are co-torsion free modules of X, invariant under Γ, and are the annihilators, respectively, of a maximal k-trivial subtorus $A \subset T$ and a maximal k-compact subtorus $T_0 \subset T$ (Example 2, I,§4). Thus if we let $Y = X(A)$, then Y may be identified with X/X_0; we will denote by π the natural projection from X to $Y = X/X_0$.

Let $\sqrt{\ }_0 = \sqrt{\ } \cap X_0$; $\sqrt{\ }_0$ is clearly a closed subsystem of $\sqrt{\ }$. Let W be the Weyl group of $\sqrt{\ }$, W_0 the Weyl group of $\sqrt{\ }_0$ (W_0 is identified with the subgroup of W generated by the reflections w_α, $\alpha \in \sqrt{\ }_0$), and define $W^\Gamma = \{w \in W | w(X_0) = X_0\}$. W_0 is clearly a subgroup of W^Γ, and in fact, W_0 is a normal subgroup of W^Γ. For, W_0 is generated by $\{w_\alpha | \alpha \in \sqrt{\ }_0\}$, and if $w \in W^\Gamma$ and $\alpha \in \sqrt{\ }_0$, then property (5) of roots implies $ww_\alpha w^{-1} = w_{w(\alpha)}$, and since $w(\alpha) \in \sqrt{\ } \cap X_0 = \sqrt{\ }_0$, it follows that $ww_\alpha w^{-1} \in W_0$. From the definition of W^Γ, we see that if $w \in W^\Gamma$, then w induces an automorphism of $X/X_0 = Y$; we denote the induced automorphism by $\pi(w)$. Thus $\pi(w\chi) = \pi(w)\pi(\chi)$.

<u>Definition</u>: Let X_0 be a co-torsion free submodule of a given module X. A linear order $>$ on X satisfying:

(7) if $\chi \notin X_0$, $\chi > 0$, and $\chi' \equiv \chi \pmod{X_0}$, then $\chi' > 0$

is called a <u>linear order adapted to</u> X_0.

It is easily verified that a linear order $>$ on a module X is characterized by a subset $X_+ = \{\chi \in X | \chi > 0\}$ satisfying (i) X_+ is closed under addition; (ii) $0 \notin X_+$, and if $\chi \neq 0$, then $\chi \in X_+$ or $-\chi \in X_+$. From this characterization, it is clear that a linear order on X adapted to X_0 induces linear orders on $Y = X/X_0$ and X_0, and conversely, given linear orders on X_0 and on Y, these uniquely determine a linear order on X adapted to X_0 which induces the given linear orders (i.e., if $\chi \notin X_0$, then define $\chi > 0$ if and only if $\pi(\chi) > 0$).

Now fix a linear order $>$ on X adapted to X_o, and let Δ be a fundamental system of $\mathfrak{r}$ with respect to this order (Δ will be called an X_o-<u>fundamental system of</u> $\mathfrak{r}$), and let Δ_o be a fundamental system of $\mathfrak{r}_o$ with respect to the induced order on X_o. We define $\bar{\mathfrak{r}} = \pi(\mathfrak{r} - \mathfrak{r}_o)$, $\bar{\Delta} = \pi(\Delta - \Delta_o)$; $\bar{\mathfrak{r}}$ is called the set of <u>restricted roots</u> of G relative to A, and $\bar{\Delta}$ is called a <u>restricted fundamental system</u> of $\mathfrak{r}$ relative to A. The following proposition lists some properties of X_o, $\mathfrak{r}$, $\mathfrak{r}_o$, Δ, Δ_o, etc. which are independent of the definition of X_o, that is, they are true for <u>any</u> co-torsion free submodule X_o of X.

<u>Proposition 2.1.1</u>: Let X_o be a co-torsion free submodule of X, and $\mathfrak{r}$ a root system in X. If $\mathfrak{r}_o$, $\bar{\mathfrak{r}}$, Δ, Δ_o, $\bar{\Delta}$ are defined as above, then

1) $\Delta_o = \Delta \cap \mathfrak{r}_o$.

2) If Δ' is another X_o-fundamental system of $\mathfrak{r}$, $\Delta_o' = \Delta' \cap \mathfrak{r}_o$, and $\bar{\Delta}' = \pi(\Delta' - \Delta_o')$, then $\Delta = \Delta'$ if and only if $\Delta_o = \Delta_o'$ and $\bar{\Delta} = \bar{\Delta}'$.

3) If $w \in W^{\Gamma}$, then $w(\Delta)$ is also an X_o-fundamental system, and the following three statements are equivalent:

(i) $w \in W_o$,

(ii) $\pi(w) = 1$,

(iii) $\pi(w)\bar{\Delta} = \bar{\Delta}$.

<u>Proof</u>:

1) It is enough to show that each $\alpha \in \mathfrak{r}_o$ is a linear combination of the α_i's in Δ_o. We may assume $\alpha > 0$; then $\alpha = \Sigma_{i=1}^{\ell} m_i \alpha_i$, $m_i \geq 0$ (Prop. 1.2.1). If there is an index j such that $\alpha_j \notin \Delta_o$ and $m_j \geq 1$, then either one has $m_j \geq 2$, or there is another index K such that $\alpha_K \notin \Delta_o$ and $m_K \geq 1$. In either case, α_j is congruent modulo X_o to a negative element, which is a contradiction.

2) ($\Leftarrow$) Let $\alpha \in \Delta$. If $\alpha \in \Delta_o = \Delta_o'$, then $\alpha >' 0$ (α is positive with respect to the order defining Δ'). If $\alpha \notin \Delta_o$, then $\pi(\alpha) \in \bar{\Delta} = \bar{\Delta}'$, so $\alpha >' 0$ in this case also. Since Δ determines $\Phi_+ = \{\alpha \in \Phi \mid \alpha > 0\}$ (Prop. 1.2.1), we see that $\Phi_+ \subset \Phi_+'$, so $\Phi_+ = \Phi_+'$ (since $\Phi = \Phi_+ \cup (-\Phi_+) = \Phi_+' \cup (-\Phi_+'))$, hence $\Delta = \Delta'$.

3) Since $w(X_o) = X_o$, it is clear that $w(\Delta)$ is an X_o-fundamental system of Φ with respect to the linear order on X defined by $X_+ = \{w(\chi) \mid \chi > 0\}$. From the definition of $\pi(w)$, it is clear that (i) => (ii), and (ii) => (iii) is trivial. Since by definition, $\pi(w)\bar{\Delta} = \pi(w(\Delta))$, (iii) => (i) results from the lemma below, and Prop. 1.2.3.

<u>Lemma 2.1.2</u>: If Δ, Δ' are X_o-fundamental systems of Φ such that $\bar{\Delta} = \overline{\Delta'}$, then there exists a unique $w_o \in W_o$ such that $\Delta' = w_o\Delta$.

<u>Proof</u>: Since Δ_o and Δ_o' are fundamental systems of Φ_o (by 1), Prop. 2.1.1) there exists a unique $w_o \in W_o$ such that $w_o\Delta_o = \Delta_o'$ (Prop. 1.2.3). Then $(w_o\Delta) \cap \Phi_o = \Delta_o'$, and $\pi(w_o\Delta) = \bar{\Delta} = \overline{\Delta'}$, so $\Delta' = w_o\Delta$ (by 2), Prop. 2.1.1).

Define $\bar{W} = \{\pi(w) \mid w \in W^\Gamma\}$. Then Prop. 2.1.1, 3) implies

<u>Corollary 2.1.3</u>: $\bar{W} \cong W^\Gamma/W_o$.

Another property which does not depend on the definition of X_o, but depends only on the fact that X_o is an annihilator of some subtorus A is the following isomorphism:

<u>Proposition 2.1.4</u>: $W^\Gamma/W_o \cong N(A)/Z(A)$, where $N(A)$ and $Z(A)$ are, respectively, the normalizer and centralizer of A in G.

The proposition is easily proved using the following lemma and the second isomorphism theorem.

Lemma 2.1.5: If X_o is a co-torsion free module of X and A is the subtorus of T annihilated by X_o, and $s \in N(T)$, then

(i) $w_s \in W^\Gamma$ if and only if $s \in N(A)$.

(ii) $w_s \in W_o$ if and only if $s \in Z(A)$.

(iii) $N(A) = (N(A) \cap N(T)) \cdot Z(A)$.

Proof: (i) By definition of w_s, ((2),(3) of §1.1) $w_s = {}^tI_s{}^{-1}$, so $w_s(X_o) = X_o$ if and only if $s \in N(A)$.

(ii) By 3) of Prop. 2.1.1, for $s \in N(A)$, one has $w_s \in W_o$ if and only if $\pi(w_s) = 1$; clearly this is true if and only if I_s is the identity on A.

(iii) Containment $\supset$ is clear. If $s \in N(A)$, then $T \subset Z(A)$ and $sTs^{-1} \subset Z(A)$, so there is an element $s_1 \in Z(A)$ such that $s_1Ts_1{}^{-1} = sTs^{-1}$ (Remark 1, I, §4.3). Clearly $s_2 = s_1{}^{-1}s \in N(T) \cap N(A)$, and $s = s_1s_2$.

If $s \in N(A) \cap N(T)$, define $\bar{w}_s = {}^t(I_s|A)^{-1}$. The isomorphisms of Coroll. 2.1.3 and Prop. 2.1.4 imply that $\pi(w_s) = \bar{w}_s$.

Although we wish to study the submodule X_o which is the annihilator of A, a maximal k-trivial torus in T, the general results obtained above raise a natural question, namely, for any co-torsion free module $X_o \subset X$, if we define the objects $\bar{\nabla}$, $\bar{\Delta}$, $\bar{W}$ in the manner described above, when is the set $\bar{\nabla}$ a root system (in a wider sense) in $Y = X/X_o$ with fundamental system $\bar{\Delta}$ and Weyl group $\bar{W}$? As we shall show, when k is perfect and when X_o is the annihilator of a maximal k-trivial torus, this is actually the case; Tits and Borel have demonstrated this result for the case of k an arbitrary field (see [8];[9]). The general question is considered by D. Schattschneider ("On restricted roots of semi-simple algebraic groups", J. Math. Soc. Japan 21 (1969), 94-115).

For the remainder of this section, we assume X_o is the annihilator in X of A, a maximal k-trivial subtorus of T (all other notation is as previously defined). In this case, a linear order > on X is adapted to X_o if and only if the following condition is satisfied:

(7') if $\chi > 0$ and $\chi \notin X_o$, then $\chi^\sigma > 0$ for all $\sigma \in \Gamma$.

For, if (7') is satisfied, and $\chi > 0$, $\chi \notin X_o$, and $\chi' \equiv \chi \pmod{X_o}$, then $\chi' = \chi + \chi_o$ for some $\chi_o \in X$, so $\sum_{\sigma\in\Gamma} \chi'^\sigma = \sum_{\sigma\in\Gamma} \chi^\sigma > 0$ (since $\sum_{\sigma\in\Gamma} \chi_o^\sigma = 0$) which implies (by (7')) that $\chi' > 0$. Conversely, if (7) is satisfied, then (7') follows, since by the definition of X_o, $\chi^\sigma \equiv \chi \pmod{X_o}$ for all $\chi \in X$.

<u>Definition</u>: A linear order on X which satisfies (7') is called a <u>Γ-linear order</u>. A fundamental system of $\mathcal{V}$ with respect to a Γ-linear order is called a Γ-<u>fundamental system of</u> $\mathcal{V}$.

<u>Proposition 2.1.6</u>: Let $\bar{\Delta} = \{\gamma_1,\ldots,\gamma_r\}$ be a restricted fundamental system of $\mathcal{V}$ (the γ_i mutually distinct). Then the elements $\gamma_1,\ldots,\gamma_r$ are linearly independent (hence rank Y = r).

<u>Proof</u>: Clearly ℓ = rank X = rank X_o + rank Y, and $\ell = \#\Delta = \#\Delta_o + \sum_{\gamma_i\in\bar{\Delta}} (\#\pi^{-1}(\gamma_i) - 1) + \#\bar{\Delta}$. For each $\gamma_i \in \bar{\Delta}$, choose a fixed $\alpha_{i_o} \in \pi^{-1}(\gamma_i)$; then $\alpha - \alpha_{i_o} \in X_o$ for all $\alpha \in \pi^{-1}(\gamma_i)$, and the set $S = \{\Delta_o, (\alpha - \alpha_{i_o}), \alpha \in \pi^{-1}(\gamma_i), \alpha \neq \alpha_{i_o}, 1 \leq i \leq r\}$ is a linearly independent subset of X_o. Thus rank $X_o \geq \#S = \#\Delta_o + \Sigma_{i=1}^r(\#\pi^{-1}(\gamma_i) - 1)$. By our first two equations, this is equivalent to rank $Y \leq \#\bar{\Delta} = r$ (which is also clear from the definitions of $\bar{\Delta}$ and Y). Therefore it suffices to show rank $X_o = \#S$. Thus, we show that S generates $X_{o\mathbb{Q}}$ over $\mathbb{Q}$. From the definition of X_o and X^Γ, it is clear that $X_{o\mathbb{Q}}$ is generated over $\mathbb{Q}$ by the set

$\{\alpha_i^\sigma - \alpha_i, \sigma \varepsilon \Gamma, \alpha_i \varepsilon \Delta\}$. If $\alpha_i \varepsilon \Delta_o$, then $\alpha_i^\sigma \varepsilon \surd \cap X_o = \surd_o$, and since Δ_o is a fundamental system of $\surd_o$, it follows that $\alpha_i^\sigma - \alpha_i \varepsilon \{\Delta_o\}_{\mathbb{Z}} \subset S_{\mathbb{Z}}$. If $\alpha_i \not\varepsilon \Delta_o$, then $\pi(\alpha_i) = \pi(\alpha_i^\sigma) = \gamma_j$ for some j, and so by lemma 2.1.6 below,

$$\alpha_i^\sigma = \alpha_\nu + \sum_{\alpha_K \varepsilon \Delta_o} c_{iK}\alpha_K \quad \text{for some } \alpha_\nu \varepsilon \pi^{-1}(\gamma_j),\ c_{iK} \varepsilon \mathbb{Z}.$$

But then $\alpha_i^\sigma - \alpha_i = (\alpha_\nu - \alpha_i) + \sum_{\alpha_K \varepsilon \Delta_o} c_{iK}\alpha_K \varepsilon S_{\mathbb{Z}}$.

<u>Lemma 2.1.6</u>: If $\pi(\alpha_i) = \gamma_j$ (where $\alpha_i \varepsilon \Delta$, $\gamma_j \varepsilon \bar{\Delta}$), and $\sigma \varepsilon \Gamma$, then $\alpha_i^\sigma = \alpha_\nu + \sum_{\alpha_K \varepsilon \Delta_o} c_{iK}\alpha_K$ for some $\alpha_\nu \varepsilon \pi^{-1}(\gamma_j)$, $c_{iK} \varepsilon \mathbb{Z}$.

<u>Proof</u>: Since Δ is a Γ-fundamental system, and $\alpha_i > 0$, we may write $\alpha_i^\sigma = \sum_{K=1}^{\ell} c_{iK}(\sigma)\alpha_K$, where $c_{iK}(\sigma) \varepsilon \mathbb{Z}$ and $c_{iK}(\sigma) \geq 0$ if $\alpha_i \not\varepsilon \Delta_o$, and $c_{iK}(\sigma) = 0$ if $\alpha_i \varepsilon \Delta_o$ and $\alpha_K \not\varepsilon \Delta_o$. By re-ordering the fundamental roots if necessary, we may assume that $\Delta - \Delta_o = \{\alpha_1,\ldots,\alpha_m\}$, $\Delta_o = \{\alpha_{m+1},\ldots,\alpha_\ell\}$. Then the matrices $(c_{ij}(\sigma))$ and $(c_{ij}(\sigma^{-1}))$ are integral, and both of the form

$$\left(\begin{array}{c|c} \geq 0 & \geq 0 \\ \hline 0 & * \end{array}\right).$$

Since the product of the two matrices is the identity matrix, it follows that the upper left submatrix is necessarily a permutation matrix, hence if $\alpha_i \not\varepsilon \Delta_o$, $\alpha_i^\sigma = \alpha_\nu + \sum_{\alpha_K \varepsilon \Delta_o} c_{iK}(\sigma)\alpha_K$, where one clearly has $\alpha_\nu \varepsilon \pi^{-1}(\gamma_j)$.

<u>Corollary 2.1.7</u>: (notation as in Proposition 2.1.6). Every $\gamma \varepsilon \bar{\surd}$ can be expressed uniquely in the form $\gamma = \pm \sum_{i=1}^{r} m_i\gamma_i$ with $m_i \varepsilon \mathbb{Z}$, $m_i \geq 0$. (This is immediate from Prop. 1.2.1.)

In Proposition 2.1.1 we have demonstrated that the group W^Γ operates on the set of all Γ-fundamental systems of $\surd$. The Galois group Γ also acts on this set, that is, if Δ is a Γ-fundamental system of $\mathcal{V}$, and $\sigma \in \Gamma$, then $\Delta^\sigma = \{\alpha^\sigma | \alpha \in \Delta\}$ is also a Γ-fundamental system of $\mathcal{V}$. For, if X_+ is the set of positive elements of X with respect to the Γ-linear order determining Δ, then $X_+^\sigma = \{\chi^\sigma | \chi \in X_+\}$ is the set of positive elements of X with respect to a new Γ-linear order, and Δ^σ is the set of simple roots with respect to this order. Since $\alpha_i \equiv \alpha_i^\sigma \pmod{X_o}$ for all $\alpha_i \in \Delta$, $\sigma \in \Gamma$, it follows that $\overline{\Delta}^\sigma = \overline{\Delta}$, hence there is a unique element $w_\sigma \in W_o$ such that $\Delta^\sigma = w_\sigma \Delta$ (Lemma 2.1.2). From this, we can define a new operation of Γ on X as follows:

$$\chi^{[\sigma]} = w_\sigma^{-1} \chi^\sigma, \quad \chi \in X, \ \sigma \in \Gamma. \tag{12}$$

It is easily verified that $\chi \to \chi^{[\sigma]}$ is an automorphism of the triple $(X, \mathcal{V}, \Delta)$ and that $\chi^{[\sigma][\tau]} = \chi^{[\sigma\tau]}$ for all $\sigma, \tau \in \Gamma$, $\chi \in X$.

Now denote $\hat{Y} = \mathrm{Hom}(Y, \mathbb{Z})$; then $\hat{Y}_{\mathbb{Q}} = \mathrm{Hom}(Y_{\mathbb{Q}}, \mathbb{Q})$, the dual of $Y_{\mathbb{Q}}$. We wish to define the analogue of Weyl chamber in the space $\hat{Y}_{\mathbb{Q}}$. Let $\langle\ ,\ \rangle$ be any $\overline{W}$-invariant metric on $Y_{\mathbb{Q}}$. Let $\overline{\Delta}$ be a restricted fundamental system of $\mathcal{V}$, and let $\gamma \in \overline{\mathcal{V}}$, and define

$$\Lambda_{\overline{\Delta}} = \{\omega \in \hat{Y}_{\mathbb{Q}} \mid \langle \gamma_i, \omega\rangle > 0 \text{ for all } \gamma_i \in \overline{\Delta}\},$$

$$H_\gamma = \{\omega \in \hat{Y}_{\mathbb{Q}} \mid \langle \gamma, \omega\rangle = 0\}.$$

(H_γ is just the hyperplane in $\hat{Y}_{\mathbb{Q}}$ defined by γ.)

<u>Proposition 2.1.8</u>: $\hat{Y}_{\mathbb{Q}} - \bigcup_{\gamma \in \overline{\mathcal{V}}} H_\gamma = \bigcup_{\overline{\Delta}:\text{r.f.s}} \Lambda_{\overline{\Delta}}$ (disjoint union).

<u>Proof</u>: Let $\overline{\Delta} = \{\gamma_1, \ldots, \gamma_r\}$. For each $\gamma \in \overline{\mathcal{V}}$, we may write $\gamma = \pm \sum n_i \gamma_i$, $n_i \geq 0$, $n_i \in \mathbb{Z}$ (Prop. 1.2.1), thus it is clear that $\Lambda_{\overline{\Delta}}$ is a component of $Y_{\mathbb{Q}} - \bigcup_{\gamma \in \overline{\mathcal{V}}} H_\gamma$. Also, by Prop. 2.1.6, if $\overline{\Delta}$ and $\overline{\Delta}'$ are distinct restricted fundamental systems, then $\Lambda_{\overline{\Delta}}, \Lambda_{\overline{\Delta}'}$ are disjoint. Let $\omega \in \hat{Y}_{\mathbb{Q}} - \bigcup_{\gamma \in \overline{\mathcal{V}}} H_\gamma$; then one can define a linear order in Y satisfying

the condition: if $\eta \in Y$ and $\langle \eta, \omega \rangle > 0$, then $\eta > 0$. Extending this order to a Γ-linear order in X, if $\bar{\Delta}$ is the restricted fundamental system determined by the order, then for every $\gamma_i \in \bar{\Delta}$, $\langle \gamma_i, \omega \rangle > 0$, hence $\omega \in \Lambda_{\bar{\Delta}}$.

2.2 Structure of parabolic subgroups ([3]§4, [9]§4)

G continues to denote a connected semi-simple algebraic group defined over k, a perfect field, and T a maximal torus of G defined over k.

In I, §4.2, §4.3, we have discussed some of the properties of Borel subgroups of G (there G was an arbitrary connected algebraic group). In addition, the following property is known (for G semi-simple).

<u>Proposition 2.2.1</u>: ([2], exposé 13)

There is a one-to-one correspondence between Borel subgroups B of G containing T and fundamental systems Δ of $\mathcal{V}$, the root system of G with respect to T. That is, given B, there exists a unique fundamental system $\Delta \subset \mathcal{V}$ such that $B = B_\Delta = T \cdot \prod_{\alpha \in \mathcal{V}_+} P_\alpha$ (a semi-direct product), where $\mathcal{V}_+$ is the set of positive roots in $\mathcal{V}$ determined by Δ , and vice versa.

We wish to demonstrate a similar correspondence between parabolic subgroups of G and subsets of fundamental systems of $\mathcal{V}$.

Let P be a parabolic subgroup of G, $P \supset B \supset T$, where $B = B_\Delta$ is a (fixed) Borel subgroup of G. We have noted (I, §4.3) that P is connected, hence the following general result holds for P.

<u>Lemma 2.2.2</u>: If H is a connected subgroup of a connected semi-simple group G, and H contains a maximal torus T of G, then H is generated by T and the set $\{P_\alpha \mid P_\alpha \subset H\}$. ([2]-12-07, Prop. 3; 13-05, Theorem 1).

Let U be the unipotent radical of P; then it is easily shown that

$U \subset B_u$, the unipotent part of B. Moreover, one sees that P is the semi-direct product $P = (S \cdot T) \cdot U$, where S is a connected semi-simple group.[10)] More precisely, if one puts $\mathcal{V}_1 = \{\alpha \in \mathcal{V} \mid P_\alpha \subset S\}$, the group $S \cdot T$ is reductive, having T as a maximal torus and $\mathcal{V}_1$ as root system. Also, if $B = T \cdot \prod_{\alpha \in \mathcal{V}_+} P_\alpha$, then clearly $U = \prod_{\alpha \in \mathcal{V}_+ - \mathcal{V}_1} P_\alpha$, therefore $\mathcal{V}_1$ can also be characterized by the property $\mathcal{V}_1 = \{\alpha \in \mathcal{V} \mid P_\alpha \text{ and } P_{-\alpha} \subset P\}$. We shall now show that $\mathcal{V}_1$ is a $\mathbb{Q}$-closed subsystem of $\mathcal{V}$.

<u>Lemma 2.2.3</u>: If α_1 is simple in $\mathcal{V}_1$, then α_1 is simple in $\mathcal{V}$.

<u>Proof</u>: Suppose $\alpha_1 = \alpha + \beta$ where $\alpha, \beta > 0$ in $\mathcal{V}$. From the definition of $c_{\alpha\beta}$, it follows that $2 = c_{\alpha_1\alpha_1} = c_{\alpha_1\alpha} + c_{\alpha_1\beta}$. Suppose $c_{\alpha_1\alpha} \geq 1$. Then $w_{\alpha_1}\alpha = \alpha - c_{\alpha_1\alpha}\alpha_1 = -\beta - (c_{\alpha_1\alpha} - 1)\alpha_1 < 0$. Since $\alpha_1 \in \mathcal{V}_1$, there exists $s_1 \in N(T) \cap S$ such that $P_{w_{\alpha_1}\alpha} = s_1 P_\alpha s_1^{-1} \subset P$. In view of $w_{\alpha_1}\alpha < 0$, this implies $w_{\alpha_1}\alpha \in \mathcal{V}_1$, hence $\alpha \in \mathcal{V}_1$. Similarly, if $c_{\alpha_1\beta} \geq 1$, then $\beta \in \mathcal{V}_1$. Since α_1 is $\mathcal{V}_1$-simple, it follows that $c_{\alpha_1\alpha}$ or $c_{\alpha_1\beta}$ is ≤ 0. Suppose $c_{\alpha_1\beta} \leq 0$. Then $c_{\alpha_1\alpha} \geq 2$ and by the argument above, $\alpha \in \mathcal{V}_1$. Since the Schwarz inequality implies $c_{\alpha_1\alpha}\, c_{\alpha\alpha_1} < 4$, it follows that $c_{\alpha\alpha_1} = 1$; but then $w_\alpha \alpha_1 = \alpha_1 - \alpha = \beta \in \mathcal{V}_1$, impossible. Similarly, if $c_{\alpha_1\alpha} \leq 0$, a contradiction results; thus α_1 must be $\mathcal{V}$-simple.

<u>Corollary 2.2.4</u>: In the above situation (where $B = B_\Delta$), $\Delta_1 = \Delta \cap \mathcal{V}_1$ is a fundamental system of $\mathcal{V}_1$.

We may reformulate this in a slightly different form as follows:

<u>Corollary 2.2.4'</u>: If Δ_1' is a fundamental system of $\mathcal{V}_1$, then there exists a fundamental system Δ' of $\mathcal{V}$ such that $\Delta_1' = \Delta' \cap \mathcal{V}_1$, and $B_{\Delta'} \subset P$.

Corollary 2.2.5: In the above situation, there exists a co-torsion free submodule $X_1 \subset X$ such that $\mathscr{V}_1 = \mathscr{V} \cap X_1$ and Δ is an X_1-fundamental system of $\mathscr{V}$.

Proof: Let $X_1 = \{\mathscr{V}_1\}_{\mathbb{Q}} \cap X$; then X_1 is co-torsion free in X. Since Δ is a basis for $X_{\mathbb{Q}}$ over $\mathbb{Q}$, one has $\Delta \cap X_1 = \Delta \cap \mathscr{V}_1 = \Delta_1$. Clearly $\Delta_1 = \Delta \cap X_1$ is a fundamental system of the root system $\mathscr{V} \cap X_1$, but by Coroll. 2.2.4, Δ_1 is a fundamental system of $\mathscr{V}_1$. Since $\mathscr{V}_1 \subset \mathscr{V} \cap X_1$, it follows that $\mathscr{V}_1 = \mathscr{V} \cap X_1$ (which implies $\mathscr{V}_1$ is $\mathbb{Q}$-closed). Now let $\Delta = (\alpha_1, \ldots, \alpha_\ell)$ and $\Delta_1 = (\alpha_{m+1}, \ldots, \alpha_\ell)$. Then, the lexicographical linear order in X with respect to the basis $(\alpha_1, \ldots, \alpha_\ell)$ (in $X_{\mathbb{Q}}$) is adapted to X_1, and Δ is clearly a fundamental system of $\mathscr{V}$ with respect to this order. Thus Δ is an X_1-fundamental system.

Remark: The above proof shows that, for a subsystem $\mathscr{V}_1$ of $\mathscr{V}$ and for a fundamental system Δ of $\mathscr{V}$, the existence of X_1 as stated in Corollary 2.2.5 is equivalent to the fact that $\Delta \cap \mathscr{V}_1$ is a fundamental system of $\mathscr{V}_1$.

Thus we have shown that if $B = B_\Delta$ is a Borel subgroup of G containing T, then any parabolic subgroup $P \supset B$ determines a subset $\Delta_1 \subset \Delta$, where $\Delta_1 = \Delta \cap \mathscr{V}_1$, and $\mathscr{V}_1 = \{\alpha \in \mathscr{V} \mid P_\alpha \text{ and } P_{-\alpha} \subset P\}$. Conversely, given a subset $\Delta_1 \subset \Delta$, the set $\mathscr{V}_1 = \{\Delta_1\}_{\mathbb{Z}} \cap \mathscr{V}$ is a $\mathbb{Q}$-closed subsystem of $\mathscr{V}$, and determines a connected semi-simple group $G(\mathscr{V}_1)$, namely the group generated by $\{P_\alpha \mid \alpha \in \mathscr{V}_1\}$, and the group $G(\mathscr{V}_1) \cdot T \cdot \prod_{\alpha \in \mathscr{V}_+ - \mathscr{V}_1} P_\alpha$ is a parabolic group of G containing B_Δ. Thus we obtain:

Proposition 2.2.6: There exists a one-to-one correspondence between parabolic subgroups P containing B_Δ and subsets Δ_1 of Δ. The parabolic subgroup corresponding to $\Delta_1 \subset \Delta$ is of the form $P_{\Delta_1} = G(\mathscr{V}_1) \cdot T_{\mathscr{V}_1} \cdot U_{\mathscr{V}_1}^{+}$, where $T_{V_1} = (V_1^{\perp})^{\circ}$ is the subtorus of T annihilated by V_1, $U_{V_1}^{+} = \prod_{\alpha \in V_+ - V_1} P_\alpha$ is the unipotent radical of P, and $T_{V_1} \cap G(V_1)$ is finite.

In addition, it is easily shown that the reductive group $G(V_1) \cdot T_{V_1}$ is just $Z(X_1^{\perp})$, the centralizer in G of the subtorus of T whose annihilator is X_1 (for $P_\alpha \in G(V_1) \Leftrightarrow \alpha \in V_1 = V \cap X_1 \Leftrightarrow \alpha(t) = 1$ for all $t \in X_1^{\perp} \Leftrightarrow P_\alpha \subset Z(X_1^{\perp})$). The conjugacy of Borel subgroups in G implies that $\{P_{\Delta_i} | \Delta_i \subset \Delta\}$ is a complete set of representatives of conjugacy classes of parabolic subgroups of G.

Suppose now that P_1 and P_2 are two parabolic subgroups of G, both containing the Borel subgroup B_Δ; using the notation above, we may write $P_i = G(V_i) \cdot T_{V_i} U_{V_i}^{+}$. From this product, it is clear that $P_1 \supset P_2 \Leftrightarrow V_1 \supset V_2 \Leftrightarrow G(V_1) \supset G(V_2) \Leftrightarrow U_{V_1}^{+} \subset U_{V_2}^{+}$.

<u>Corollary 2.2.6</u>: If U is the unipotent radical of a parabolic subgroup P of G, then P = N(U), the normalizer of U in G. Thus the correspondence $P = N(U) \longleftrightarrow U$ is a one-to-one correspondence between parabolic subgroups of G and their unipotent radicals.

<u>Proof</u>: Since U is normal in P, we have $P \subset N(U)$, hence N(U) is a parabolic subgroup of G. Thus if U' is the unipotent radical of N(U), the relationship above shows $U' \subset U$; but clearly $U \subset U'$, hence $U = U'$, and hence $P = N(U)$.

<u>Remark</u>: In general, each connected unipotent subgroup of G does not correspond to a parabolic subgroup of G; only those connected unipotent subgroups normalized by a maximal torus of G can be unipotent radicals of parabolic subgroups.

We now wish to investigate the structure of parabolic subgroups of

G which are <u>defined</u> <u>over</u> k and contain T (which is also defined over k). (We use notation as defined in §2.1.) Suppose $P \supset B_\Delta \supset T$; write $P = G(\sqrt{}_1)\cdot T_{\sqrt{}_1}\cdot U_{\sqrt{}_1}^{+}$. Now P is defined over k if and only if

(i) $$P^\sigma = P \quad \text{for all } \sigma \in \mathrm{Gal}(\bar{k}/k).$$

Since $G(\sqrt{}_1)$ and $U_{\sqrt{}_1}^{+}$ are uniquely determined by P, and T is defined over k, (i) is equivalent to

(ii) $$G(\sqrt{}_1)^\sigma = G(\sqrt{}_1) \text{ and } (U_{\sqrt{}_1}^{+})^\sigma = U_{\sqrt{}_1}^{+} \text{ for all } \sigma \in \mathrm{Gal}(\bar{k}/k).$$

From the definitions of $G(\sqrt{}_1)$ and $U_{\sqrt{}_1}^{+}$, and since $(P_\alpha)^\sigma = P_{\alpha^\sigma}$ for all $\sigma \in \mathrm{Gal}(\bar{k}/k)$, it is clear that (ii) is equivalent to

(iii) $$\sqrt{}_1^{\,\sigma} = \sqrt{}_1, \text{ and } (\sqrt{}_+ - \sqrt{}_1)^\sigma = \sqrt{}_+ - \sqrt{}_1, \quad \text{for all } \sigma \in \mathrm{Gal}(\bar{k}/k).$$

Since every $\chi \in X$ is defined over K (the Galois splitting field of T), the action of $\mathrm{Gal}(\bar{k}/k)$ on X is essentially the action of $\Gamma = \mathrm{Gal}(K/k)$, so that we may replace $\mathrm{Gal}(\bar{k}/k)$ by Γ in (iii). We now show that (iii) is equivalent to

(iv) X_1 can be taken so that $X_1 \supset X_o$ (X_1 as in Corollary 2.2.5).

If $X_1 \supset X_o$, then $X_1/X_o \subset X/X_o$; since Γ leaves X_o invariant and Γ acts trivially on X/X_o, it follows that $X_1^\sigma = X_1$, hence $\sqrt{}_1^{\,\sigma} = \sqrt{}_1$ for all $\sigma \in \Gamma$. Moreover (in the situation of Corollary 2.2.5) $\sqrt{}_+$ is determined by an order on X adapted to X_1; therefore, if $\alpha \in \sqrt{}_+ - \sqrt{}_1$, then $\alpha^\sigma \equiv \alpha \pmod{X_o}$ implies $\alpha^\sigma \equiv \alpha \pmod{X_1}$, hence $\alpha^\sigma \in \sqrt{}_+ - \sqrt{}_1$, for all $\sigma \in \Gamma$. Conversely, suppose $\sqrt{}_1^{\,\sigma} = \sqrt{}_1$ and $(\sqrt{}_+ - \sqrt{}_1)^\sigma = \sqrt{}_+ - \sqrt{}_1$ for all $\sigma \in \Gamma$. Put $X_1 = \{\sqrt{}_1\}_{\mathbb{Q}} \cap X$, and $X_1' = X_1 + X_o$. Since X_1 and X_o are cotorsion free submodules of X, so is X_1'; in order to show that X_1' may replace X_1 in Corollary 2.2.5, we only need to show that $\sqrt{}_1 = \sqrt{} \cap X_1'$ (the proof that Δ is an X_1' fundamental system is the same as given in the corollary). Let $\alpha \in \mathcal{V}_+ \cap X_1'$, write $\alpha = \chi_1 + \chi_o$, $\chi_1 \in X_1$, $\chi_o \in X_o$. Since $\Sigma_{\sigma \in \Gamma} \chi_o^{\sigma} = 0$, and $X_1^{\sigma} = X_1$, one has $\Sigma_{\sigma \in \Gamma} \alpha^{\sigma} = \Sigma_{\sigma \in \Gamma} \chi_1^{\sigma} \in X_1$, hence $\alpha \in \mathcal{V}_1$ (for if $\alpha \notin \mathcal{V}_1$, then $\alpha^{\sigma} \in \mathcal{V}_+ - \mathcal{V}_1$, all $\sigma \in \Gamma$, and $\mathcal{V}_+$ determined by an order adapted to X_1 implies $\Sigma_{\sigma \in \Gamma} \alpha^{\sigma} \notin X_1$).

In example 2, chapter I, §4.4, we have seen that a co-torsion free submodule X_1 of X contains X_o if and only if the torus $A_1 = X_1^{\perp}$ is k-trivial (and contained in A). Thus (iv) is equivalent to:

(v) There exists a k-trivial torus A_1 in T such that the reductive part of P is equal to $Z(A_1)$, and Δ is an X_1-fundamental system of $\mathcal{V}$, where $X_1 = A_1^{\perp}$.

A linear order in X adapted to $X_1 \supset X_o$ may easily be modified (in X_1) so that it is also adapted to X_o; this modification does not affect the relation $P \supset B_{\Delta}$. Thus, for a parabolic subgroup P defined over k and containing T, one can find a Γ-fundamental system Δ of $\mathcal{V}$ such that $P \supset B_{\Delta}$, and $\Delta \supset \Delta_1 \supset \Delta_o$. Applying π to these subsets of Δ, it is easily verified that the following correspondence is one-to-one:

$$\{P_{\Delta_1/k} \mid P_{\Delta_1} \supset B_{\Delta}\} \longleftrightarrow \{\bar{\Delta}_1 \subset \bar{\Delta}\},$$

where Δ is a fixed Γ-fundamental system of $\mathcal{V}$, and $\Delta_1 = \pi^{-1}(\bar{\Delta}_1 \cup \{0\}) \cap \Delta$. (In fact, by Prop. 2.1.6, for any subset $\bar{\Delta}_1$ of $\bar{\Delta}$, one can modify the given Γ-linear order (without changing Δ) in such a way that the induced linear order in $Y = X/X_o$ is adapted to $Y_1 = Y \cap \{\bar{\Delta}_1\}_{\mathbb{Q}}$. Putting $X_1 = \pi^{-1}(Y_1)$, one sees that X_1 is a co-torsion free submodule of X such that $\Delta_1 = \Delta \cap X_1$, Δ is an X_1-fundamental system, and $X_1 \supset X_o$. Thus P_{Δ_1} is a k-parabolic subgroup). The above correspondence implies that the conjugacy classes of parabolic subgroups of G which are defined over k and contain T are in one-to-one correspondence with the subsets of $\bar{\Delta}$.

Another immediate consequence of the above correspondence is that a minimal parabolic subgroup P defined over k and containing T must correspond to $\emptyset \subset \bar{\Delta}$, hence $P = P_{\Delta_0}$. From (v), it follows that $P = Z(A) \cdot U$, where A is the (unique) maximal k-trivial subtorus of T, and $U = \prod_{\pi(\alpha)>0} P_\alpha$. From Prop. 2.2.6, it follows that U is maximal in the set of connected unipotent subgroups of G which are defined over k and which correspond to parabolic subgroups defined over k and containing T. It can also be shown that U is a maximal element in the set of all connected unipotent subgroups of G which are defined over k and normalized by Z(A) (see [8]).

Lemma 2.2.7: If U is a maximal element in the set of unipotent connected subgroups of G which are defined over k, then there exists a maximal k-trivial torus A in G such that U is normalized by Z(A). (For char. k = 0, see [3], 206-20/21, and for char. k = p, Tits has announced the result, [9].)

Proposition 2.2.8: Let T be a maximal torus defined over k in G, containing a maximal k-trivial torus A in $G^{(*)}$, and let U be a maximal element in the set of all connected unipotent subgroups of G which are defined over k and normalized by Z(A). Then P = Z(A)U is a minimal parabolic subgroup of G defined over k.

(This follows immediately from the lemma and our remarks above.)

(*)Remark: The existence of T as stated in Prop. 2.2.8 is proved as follows: Z(A) is connected ([2]-6-14), and defined over k, hence contains a maximal torus T defined over k (I, §4.2, Remark 2); clearly, one then has $T \supset A$.

Proposition 2.2.9: The following three conditions on G are equivalent:

(i) G has no k-trivial torus (i.e., $A = \{1\}$).

(ii) G has no connected unipotent subgroup defined over k (i.e., $U = \{1\}$).

(iii) $N(A)_k = Z(A)_k$.

Proof:

(i) => (ii). If G contains a connected unipotent subgroup $U' \neq \{1\}$, defined over k, then there is a maximal such subgroup U, and by Proposition 2.2.8, U determines a minimal parabolic subgroup of G, defined over k, and containing a maximal k-trivial torus $A \neq \{1\}$ of G.

(ii) => (i). If G contains a k-trivial torus $A' \neq \{1\}$, then $A \neq \{1\}$ (since maximal k-trivial tori of G are conjugate, coroll. 4.4.2). But $A \neq \{1\}$ implies $X_0 = A^{\perp} \subsetneqq X$, and since a minimal parabolic subgroup of G defined over k is of the form $P = Z(A)U$, where $U = \prod_{\alpha \in \mathcal{V}_+ - \mathcal{V}_0} P_\alpha$, we see that $U \neq \{1\}$.

(i) => (iii) is trivial.

(iii) => (i). If $A \neq \{1\}$, then $U = \prod_{\pi(\alpha)>0} P_\alpha \neq \{1\}$; set $U' = \prod_{\pi(\alpha)<0} P_\alpha$. Then AU and AU' are two distinct k-Borel subgroups in G (Prop. 4.4.1), so there is an element $g \in G_k$ satisfying $gAUg^{-1} = AU'$ (Prop. 4.4.2). Since gAg^{-1} and A are both maximal k-trivial tori in AU', there is an element $h \in (AU')_k$ satisfying $hAh^{-1} = gAg^{-1}$ (coroll. 4.4.2). Clearly $h^{-1}g \in N(A)_k$ and $h^{-1}gUg^{-1}h = U'$, so $h^{-1}g \notin Z(A)$ (since Z(A) normalizes U).

Remark: Conditions (i) and (ii) are equivalent to: G has no k-solvable subgroup (Prop. 4.4.1). Thus G is k-compact if any of the three conditions in the proposition are satisfied.

2.3 The set $\bar{\mathcal{V}}$ is a root system in a wider sense ([8]).

Throughout this section, we continue to use the same notations as in §2.1,2.2; G is a connected semi-simple algebraic group defined over a perfect field k, T a maximal torus defined over k containing A, a maximal k-trivial torus of G, and $\sqrt{\Gamma}$ the root system of G with respect to T.

We will prove in several steps the following

<u>Theorem 2.3.1</u>: If Δ is a Γ-fundamental system of $\sqrt{\Gamma}$, then $\bar{\sqrt{\Gamma}}$ is a root system in a wider sense, having $\bar{\Delta}$ as a fundamental system, and $\bar{W}$ is the Weyl group of $\bar{\sqrt{\Gamma}}$.

In the arguments below, we assume $A \neq \{1\}$, since otherwise, $\bar{\sqrt{\Gamma}} = \phi$ and $\bar{W} = \{1\}$, so there is nothing to prove. Of course, if $A = T$, the theorem simply restates the well-known results in [2].

1° $\bar{W}$ operates simply transitively on the set of all restricted fundamental systems of $\sqrt{\Gamma}$.

In Prop. 2.1.1, we proved that if $\bar{w} \in \bar{W}$ and $\bar{\Delta}$ is a restricted fundamental system of $\sqrt{\Gamma}$, then $\bar{w}(\bar{\Delta})$ is also a restricted fundamental system of $\sqrt{\Gamma}$, and $\bar{w}(\bar{\Delta}) = \bar{\Delta} \Leftrightarrow \bar{w} = 1$. Thus we only need to show $\bar{W}$ acts transitively on the set of all restricted fundamental systems of $\sqrt{\Gamma}$. In the last section, we saw that there was a one-to-one correspondence between restricted fundamental systems of $\sqrt{\Gamma}$ and maximal, connected unipotent subgroups of G defined over k and normalized by Z(A):

$$\bar{\Delta} \longleftrightarrow U_{\bar{\Delta}} = \prod_{\pi(\alpha)>0} P_{\alpha}.$$

If $s \in N(A)_k$, then s determines an element $\bar{w}_s \in \bar{W}$ (coroll. 2.1.3, Prop. 2.1.4), and it is easily checked that

$$\bar{w}_s(\bar{\Delta}) \longleftrightarrow s\, U_{\bar{\Delta}}\, s^{-1} = \prod_{\pi(\alpha)>0} P_{w_s\alpha}.$$

It follows from Prop. 4.4.1 and Prop. 4.4.2 that the elements of the set $\{U_{\bar{\Delta}}\}_{\bar{\Delta}:r.f.s.}$ are all conjugate by elements in $N(A)_k$. Since $N(A)_k/Z(A)_k$ may be considered as a subgroup of $\bar{W}$ in a natural manner (Coroll. 2.1.3,

Prop. 2.1.4) and since (by the one-to-one correspondence above) this subgroup acts transitively on $\{\bar{\Delta}\}_{\bar{\Delta}:\text{r.f.s.}}$, while $\bar{W}$ acts simply on $\{\bar{\Delta}\}$, it follows that $N(A)_k/Z(A)_k = \bar{W}$, hence the result.

We note that in the proof we have obtained:

1°.1 Every coset in $N(A)/Z(A)$ has a k-rational representative.

1°.2 $\bar{W}$ acts simply transitively on the set $\{\Lambda_{\bar{\Delta}}\}_{\bar{\Delta}:\text{r.f.s.}}$ (see the end of §2.1; clearly $\bar{w}(\Lambda_{\bar{\Delta}}) = \Lambda_{\bar{w}(\bar{\Delta})}$).

1°.3 W^{Γ} acts simply transitively on the set of all Γ-fundamental systems of $\mathcal{V}$ (see Prop. 2.1.1 and Lemma 2.1.2).

2° For each $\gamma \in \bar{\mathcal{V}}$, there exists a "reflection" $\bar{w}_\gamma \in \bar{W}$, that is, $\bar{w}_\gamma$ satisfies $\bar{w}_\gamma \neq 1$, $\bar{w}_\gamma^{\,2} = 1$ and $\bar{w}_\gamma | H_\gamma$ = identity.

For each $\gamma \in \bar{\mathcal{V}}$, define a subtorus Q_γ of codimension one in A by: $Q_\gamma = \{t \in A | \gamma(t) = 1\}_0$. If $s \in N(A)$, then the definitions of H_γ and Q_γ imply that $s \in Z(Q_\gamma) \Leftrightarrow \bar{w}_s | H_\gamma$ = identity. Also, $Z(Q_\gamma) \supsetneqq Z(A)$ is clear, since if $\alpha \in \mathcal{V}$ satisfies $\pi(\alpha) = \gamma$, then $P_\alpha \subset Z(Q_\gamma)$, but $P_\alpha \not\subset Z(A)$. This implies that the semi-simple part Z' of the reductive group $Z(Q_\gamma)$ is not k-compact (for if $A \cap Z' = \{1\}$ then $Z(A) = Z(Q_\gamma)$). Thus Prop. 2.2.9 and the Remark following it imply that $N(A) \cap Z(Q_\gamma) \supsetneqq Z(A)$. Taking $s \in N(A) \cap Z(Q_\gamma)$, $s \notin Z(A)$, we see that $\bar{w}_s | H_\gamma$ = identity, and $\bar{w}_s \neq 1$ (Lemma 2.1.5). We show $\bar{w}_s^{\,2} = 1$, hence we may take $\bar{w}_\gamma = \bar{w}_s$. From Prop. 2.1.8, and 1°.2, it follows that

$$\hat{Y}_Q - \bigcup_{\gamma \in \bar{\mathcal{V}}} H_\gamma = \bigcup_{\bar{w} \in \bar{W}} \bar{w}\Lambda_{\bar{\Delta}},$$

hence if $\Lambda_{\bar{\Delta}}$ is taken so that H_γ is one of its walls, then $\bar{w}_s\Lambda_{\bar{\Delta}}$ also has H_γ as one of its walls, and $\bar{w}_s\Lambda_{\bar{\Delta}} \neq \Lambda_{\bar{\Delta}}$. Since $\bar{w}_s^{\,2}\Lambda_{\bar{\Delta}} \neq \bar{w}_s\Lambda_{\bar{\Delta}}$, and H_γ is a wall of each of these chambers, it follows that $\bar{w}_s^{\,2}\Lambda_{\bar{\Delta}} = \Lambda_{\bar{\Delta}}$, hence $\bar{w}_s^{\,2} = 1$ (only two chambers may have a wall in common).

<u>Remark 2°.1</u>: Since $\bar{w}_\gamma$ is a reflection, it follows that for any $\bar{W}$-invariant metric $\langle\ ,\ \rangle$ on $Y_{\mathbb{Q}}$, $\bar{w}_\gamma$ is just the mapping

$$\bar{w}_\gamma : \eta \to \eta - \frac{2\langle\gamma,\eta\rangle}{\langle\gamma,\gamma\rangle}\gamma, \qquad \eta \in Y.$$

3° For each $\gamma \in \bar{\mathcal{V}}$, $\eta \in Y$, one has $\frac{2\langle\gamma,\eta\rangle}{\langle\gamma,\gamma\rangle} \in \mathbb{Z}$.

Let $(\ ,\ )$ be a W-invariant metric on $X_{\mathbb{Q}}$. For any $\alpha \in \mathcal{V}$, $\chi \in X$, one has $\frac{2(\alpha,\chi)}{(\alpha,\alpha)} \in \mathbb{Z}$, so $w_\alpha\chi - \chi \in \{\mathcal{V}\}_{\mathbb{Z}}$, and hence for any $w \in W$, $w\chi - \chi \in \{\mathcal{V}\}_{\mathbb{Z}}$ (see §1.1). In particular, taking $w \in W^\Gamma$, and applying π, this implies that for every $\bar{w} \in \bar{W}$, $\eta \in Y$, $\bar{w}\eta - \eta \in \{\bar{\mathcal{V}}\}_{\mathbb{Z}}$. Thus $\bar{w}_\gamma\eta - \eta =$ $-2\frac{\langle\gamma,\eta\rangle}{\langle\gamma,\gamma\rangle}\gamma \in \{\bar{\mathcal{V}}\}_{\mathbb{Z}}$. If $\frac{1}{2}\gamma \notin \bar{\mathcal{V}}$, then $\frac{2\langle\gamma,\eta\rangle}{\langle\gamma,\gamma\rangle} \in \mathbb{Z}$ (all elements of $\bar{\Delta}$, for instance, satisfy this, by Coroll. 2.1.7). To show that $\frac{2\langle\gamma,\eta\rangle}{\langle\gamma,\gamma\rangle} \in \mathbb{Z}$ for any $\gamma \in \bar{\mathcal{V}}$, we reduce the problem to the one-dimensional case, i.e., where $\bar{\Delta} = \{\gamma\}$. Thus, let $\gamma \in \bar{\mathcal{V}}$, and define $\mathcal{V}_\gamma = \{\alpha \in \mathcal{V} \mid \pi(\alpha) \in \{\gamma\}_{\mathbb{Z}}\}$. Then $\mathcal{V}_\gamma$ is clearly a closed subsystem of $\mathcal{V}$ which is Γ-invariant. The connected semi-simple group $G(\mathcal{V}_\gamma)$ (generated by $\{P_\alpha \mid \alpha \in \mathcal{V}_\gamma\}$) is defined over k, hence $G(\mathcal{V}_\gamma)T$ is a connected reductive group defined over k, having A as a maximal k-trivial torus. Clearly $\mathcal{V}_\gamma$ is the root system of $G(\mathcal{V}_\gamma)T$, and $\bar{\Delta} = \{\gamma\}$ is a restricted fundamental system of $\mathcal{V}_\gamma$. All of our arguments in §2.1, 2.2, 2.3 hold (with slight modification) for G a connected reductive group defined over k, hence, applying our argument at the beginning of 3° to the fundamental root γ of $\mathcal{V}_\gamma$, we see that $\frac{2\langle\gamma,\eta\rangle}{\langle\gamma,\gamma\rangle} \in \mathbb{Z}$.

4° $\bar{W}$ is generated by $\{\bar{w}_{\gamma_i} \mid \gamma_i \in \bar{\Delta}\}$.

The proof of this fact is the traditional one, that is, one argues using the chambers $\Lambda_{\bar{\Delta}}$. By 2°, $\bar{W}$ contains reflections with respect to every hyperplane H_γ in $\hat{Y}_{\mathbb{Q}}$, $\gamma \in \surd$, and $\hat{Y}_{\mathbb{Q}} = \bigcup_{\gamma \in \surd} H_\gamma \cup \bigcup_{\bar{w} \in \bar{W}} \bar{w}\Lambda_{\bar{\Delta}}$. Let $\bar{W}'$ be the subgroup of $\bar{W}$ generated by $\bar{w}_{\gamma_i}$, $\gamma_i \in \bar{\Delta}$. By a topological argument, one can show $\bigcup_{\bar{w} \in \bar{W}'} \bar{w}\Lambda_{\bar{\Delta}} = \bigcup_{\bar{w} \in \bar{W}} \bar{w}\Lambda_{\bar{\Delta}}$, and since $\bar{W}$ acts simply transitively on the set of Weyl chambers, it follows that $\bar{W}' = \bar{W}$. (An alternate argument appears in [2], exposé 11.)

5° Since properties (i),(ii),(iii),(iv) of §1.1 are satisfied by $(Y,\bar{\surd},\bar{W})$, we have shown that $\bar{\surd}$ is a root system in a wider sense, having $\bar{W}$ as Weyl group. Corollary 2.1.7 shows that $\bar{\Delta}$ is a fundamental system of $\bar{\surd}$ (Prop. 1.2.1 and Remark following).

2.4 The fundamental theorem of classification ([8], [10])[10a)]

All notations remain the same in this section; $G \supset T \supset A$, $X, \surd, X_0$, $\surd_0$, etc. are as in §2.3. In addition, define $G_0 = G(\surd_0)$, the connected semi-simple subgroup of G generated by $\{P_\alpha \mid \alpha \in \surd_0\}$, $T^0 = T \cap G(\surd_0)$, $X^0 = X(T^0)$. (In the following, X^0 will be identified with a submodule of $X_{\mathbb{Q}}$.)

G_0 is just the semi-simple part of Z(A), and clearly G_0 is defined over k. By Prop. 2.2.9, G_0 is k-compact. The group T^0 is a maximal torus of G_0 ([2], exposé 17) which is defined over k, and is k-compact. Since the k-conjugacy class of A is uniquely determined, and G_0 is the semi-simple part of Z(A), it follows that G_0 is uniquely determined (up to k-isomorphism) by the k-isomorphism class of G. Thus we will call G_0 the k-compact kernel of G.

Now fix Δ, a Γ-fundamental system of $\surd$, and let $\Delta_0 = \Delta \cap X_0$. In equation (12), §2.1, we have defined an action of Γ on Δ; we will denote this action by $[\sigma]$.

<u>Definition</u>. $(X, \Delta, \Delta_0, [\sigma])$ is called the Γ-diagram of G. (In example 1 below, we show that this 4-tuple is essentially unique, and does not depend on the choice of T or Δ, hence calling it "the" Γ-diagram is justified.)

In the Γ-diagram of G, the pair (X, Δ) is just the ordinary Dynkin diagram of G, Δ_0 indicates the fundamental roots annihilating A, and $[\sigma]$ indicates the action of Γ on Δ. The Γ-diagram may be illustrated by coloring black the vertices of the ordinary Dynkin diagram which represent roots in Δ_0, and indicating the action of Γ by arrows; for example,

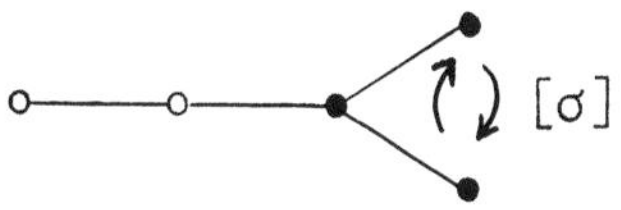

We now show that the Γ-diagram of G is uniquely determined by the k-isomorphism class of G.

Let G' be another connected semi-simple algebraic group defined over k, and T'/k, A', X', X_0', etc. denote the corresponding objects to those already defined for G. Suppose $h:(G,T,A) \to (G',T',A')$ is a K-isomorphism of G onto G' (which maps $T \to T'$, $A \to A'$). Define $h^* = {}^t(h|T)^{-1}$; then $h^*:(X, \surd, X_0) \to (X', \surd', X_0')$ is an isomorphism. Since Δ is a Γ-fundamental system of $\surd$, $h^*(\Delta)$ is a Γ-fundamental system of $\surd'$ (since it is defined by a linear order adapted to X_0'). If Δ' is a fixed Γ-fundamental system of $\surd'$, then there is a unique element $w' \in W'^{\Gamma}$ such that $w'h^*(\Delta) = \Delta'$ (§2.3, 1°.3). If we put $h^{[*]} = w'h^*$, then $h^{[*]}$ is an isomorphism mapping $(X, \surd, X_0, \Delta, \Delta_0)$ onto $(X', \surd', X_0', \Delta', \Delta_0')$. If we denote by $[\sigma]'$ the action of Γ on X' defined by (12), then $h^{[*]}$ satisfies the relationship $[\sigma]' = h^{[*]}[\sigma]h^{[*]-1}$. For, by (12), $\chi'^{[\sigma]'} = w_\sigma'^{-1} \chi'^{\sigma}$ for all $\chi' \in X'$, where $w_\sigma'(\Delta') = \Delta'^{\sigma}$, so

$w_\sigma{}' \; h^{[*]}(\Delta) = h^{[*]\sigma}(\Delta^\sigma) = h^{[*]\sigma}(w_\sigma(\Delta))$, which implies $w_\sigma{}' =$ $h^{[*]\sigma} w_\sigma h^{[*]-1}$. Thus $\chi'^{[\sigma]'} = h^{[*]} w_\sigma{}^{-1}(h^{[*]-1}\chi')^\sigma = h^{[*]}(h^{[*]-1}\chi')^{[\sigma]}$, and the stated relationship follows.

Definition: A congruence ρ of the Γ-diagram $(X,\Delta,\Delta_o,[\sigma])$ of G onto the Γ-diagram $(X',\Delta',\Delta_o',[\sigma]')$ of G' is an isomorphism which maps $(X,\Delta,\Delta_o) \to (X',\Delta',\Delta_o')$, and satisfies $[\sigma]' = \rho[\sigma]\rho^{-1}$.

We have just shown that any isomorphism $h:(G,T,A) \to (G',T',A')$ defines a congruence $h^{[*]}$ of the Γ-diagram of G onto the Γ-diagram of G'. We will call $h^{[*]}$ the congruence associated to h.

Remark: In the special case that G is k-compact ($G = G_o$), one has $\Delta = \Delta_o$, so the Γ-diagram of G may be written $(X,\Delta_o,[\sigma])$. Applying this to the k-compact kernels G_o, G_o' of G, G' it is easily seen that a congruence $\rho:(X,\Delta,\Delta_o,[\sigma]) \to (X',\Delta',\Delta_o',[\sigma]')$ induces a congruence $\rho_o:(X^o,\Delta_o,[\sigma]|X^o) \to (X'^o,\Delta'_o,[\sigma]'|X^{o\prime})$ of the Γ-diagram of G_o onto the Γ-diagram of G_o'; ρ_o is called the restriction of ρ to $(X^o,\Delta_o,[\sigma]|X^o)$. If ρ is the congruence associated to an isomorphism $h:G \to G'$ ($\rho = h^{[*]}$), then ρ_o is associated to $h_o = h|G_o$.

Example 1.

Take $G = G'$, A, A' two maximal k-trivial tori of G and T, T' any two maximal tori of G defined over k, splitting over K, and containing A and A' respectively. There is an element $g_1 \in G_k$ such that $g_1Ag_1{}^{-1} = A'$; then $g_1Tg_1{}^{-1}$ and T' are both maximal K-trivial tori in Z(A'), hence there is an element $g_2 \in Z(A)_K$ satisfying $g_2g_1Tg_1{}^{-1}g_2{}^{-1} = T'$ (Coroll. 4.4.2). Thus if $g = g_2g_1$, then the K-inner automorphism I_g of G maps $T \to T'$ and $A \to A'$, so that I_g determines a congruence $I_g{}^{[*]}$ of the Γ-diagram $(X,\Delta,\Delta_o,[\sigma])$ of G with respect to T onto the Γ-diagram $(X',\Delta',\Delta_o',[\sigma]')$ of G with respect to T'. Thus the Γ-diagram of G does not depend (up to congruence) on the choice of T and A, or Δ.

Example 2.

Let G, G' be connected semi-simple groups defined over k, and suppose $h:G \to G'$ is a k-isomorphism. Then h maps a maximal k-trivial torus A of G onto a maximal k-trivial torus A' of G', and a maximal torus T/k of G containing A onto a maximal torus T'/k of G' containing A'. Thus there is a congruence $h^{[*]}$ associated to h which maps the Γ-diagram of G onto the Γ-diagram of G'. This shows the Γ-diagram of G is uniquely determined (up to congruence) by the k-isomorphism class of G.

We have demonstrated above that the k-isomorphism class of G uniquely determines G_o, the k-compact kernel of G, and $(X,\sqrt{\ },\Delta_o,[\sigma])$, the Γ-diagram of G. The theorem which follows states that conversely, the k-compact kernel and the Γ-diagram of G determine G up to k-isomorphism. Thus the theorem reduces the problem of classifying connected semi-simple algebraic groups defined over k to the following two problems:

(i) classify all k-compact semi-simple algebraic groups;

(ii) classify all Γ-diagrams which actually belong to connected semi-simple algebraic groups defined over k.

Theorem 2.4.1: Let G, G' be connected semi-simple algebraic groups defined over k. Let T, A, X, G_o, T^o, etc., T', A', X', G_o', $T^{o\prime}$, etc. be as defined above, and corresponding to G and G', respectively. If the following conditions are satisfied, namely,

(i) there exists a congruence $\rho:(X,\Delta,\Delta_o,[\sigma]) \to (X',\Delta',\Delta_o',[\sigma]')$ of the Γ-diagram of G onto the Γ-diagram of G',

(ii) there exists a k-isomorphism $h_o:(G_o,T^o) \to (G_o',T^{o\prime})$ such that the restriction ρ_o of ρ to $(X^o,\Delta_o,[\sigma]|X^o)$ is associated to h_o

(i.e., $h_o^{[*]} = \rho_o$),

then there exists a k-isomorphism $h:(G,T) \to (G',T')$ extending h_o such that ρ is associated to h.

The theorem is proved in several steps; we begin by investigating more closely isomorphisms of semi-simple groups.

1° Let G, T be as in the theorem, and for each $\alpha \in \sqrt{\ }$, fix an isomorphism $x_\alpha : \mathfrak{G}_a \to P_\alpha$ satisfying (1), §1.1. We may assume the splitting field K of T is taken large enough so that x_α is defined over K for all $\alpha \in \sqrt{\ }$. (By the remark below, K need not be extended at all; in fact K may be taken as the smallest Galois extension of k which splits T.) Now x_α is unique up to a scalar multiple, hence if $\sigma \in \Gamma$ is applied to both sides of equation (1), one obtains

$$x_\alpha^\sigma(\xi) = x_{\alpha^\sigma}(\xi_{\alpha,\sigma}\xi) \tag{13}$$

for all $\xi \in \Omega$, with $\xi_{\alpha,\sigma} \in K^*$. From (13), we see that if $\tau \in \Gamma$, then $x_\alpha^{\sigma\tau}(\xi) = x_{\alpha^\sigma}^{\tau}(\xi_{\alpha,\sigma}^{\tau}\xi) = x_{\alpha^{\sigma\tau}}(\xi_{\alpha,\sigma}^{\tau}\xi_{\alpha^\sigma,\tau}\xi)$, hence the system $\{\xi_{\alpha,\sigma}\}$ satisfies the condition

$$\xi_{\alpha,\sigma\tau} = \xi_{\alpha,\sigma}^{\tau}\,\xi_{\alpha^\sigma,\tau}\,. \tag{14}$$

<u>Remark</u>: If $\alpha \in \sqrt{\ }$, and α is defined over K', where $k \subset K' \subset K$, then one can take x_α to be defined over K'. To see this, put $\Gamma_\alpha = \{\sigma \in \Gamma | \alpha^\sigma = \alpha\}$, and let $K_\alpha \subset K$ be the fixed field of Γ_α. By (14), the system $\{\xi_{\alpha,\sigma}\}_{\sigma \in \Gamma_\alpha}$ is a one cocycle of Γ_α in K*, so by Hilbert's Theorem 90, it follows that there is an element $\eta_\alpha \in K^*$ such that $\xi_{\alpha,\sigma} = \eta_\alpha^\sigma \eta_\alpha^{-1}$. But then if $\bar{x}_\alpha$ is defined by $\bar{x}_\alpha(\xi) = x_\alpha(\eta_\alpha^{-1}\xi)$, $\bar{x}_\alpha$ satisfies (1), and $\bar{x}_\alpha^\sigma = \bar{x}_\alpha$ for all $\sigma \in \Gamma_\alpha$, hence $\bar{x}_\alpha$ is defined over K_α.

2° Let G, G', T, T', etc. be as in the theorem, and assume K is a splitting field for T and T'. For each $\alpha \in \mathcal{V}$, $\alpha' \in \mathcal{V}'$, fix x_α, $x'_{\alpha'}$ K-isomorphisms satisfying (1), and let $\{\xi_{\alpha,\sigma}\}$, $\{\xi'_{\alpha',\sigma}\}$ be systems defined by (13) (by the x_α and $x'_{\alpha'}$ respectively). Suppose there is a K-isomorphism $h:(G,T) \to (G',T')$. Then h^* $(= {}^t(h|T)^{-1})$ is an isomorphism of $(X,\mathcal{V}) \to (X',\mathcal{V}')$ satisfying $h(P_\alpha) = P'_{h^*(\alpha)}$ (this follows easily from (8), or directly from (1), applying h to both sides). Since $h \circ x_\alpha$ is an isomorphism from $\mathcal{G}_a$ to $P'_{h^*(\alpha)}$, the 'uniqueness' of $x'_{h^*(\alpha)}$ implies

$$(15) \qquad h(x_\alpha(\xi)) = x'_{h^*(\alpha)}(\eta_\alpha \xi)$$

where $\eta_\alpha \in K^*$, and the η_α are uniquely determined by h (since the x_α and $x'_{\alpha'}$ are fixed). Since G is generated by T and $\{P_\alpha | \alpha \in \mathcal{V}\}$, equation (15) shows that h is uniquely determined by h^* and $\{\eta_\alpha | \alpha \in \mathcal{V}\}$. Thus we will write $h \longleftrightarrow \{h^*, \eta_\alpha(\alpha \in \mathcal{V})\}$.

Lemma 1: Let the notations be as above.

(i) If $\sigma \in \Gamma$, then $h^\sigma \leftrightarrow \{h^{*\sigma}, \eta^\sigma_{\alpha^{\sigma^{-1}}} \dfrac{\xi'_{h^*(\alpha^{\sigma^{-1}}),\sigma}}{\xi_{\alpha^{\sigma^{-1}},\sigma}}\}$.

(ii) If G", T" satisfy the same conditions as G, T in the theorem, and if h' is a K-isomorphism of $(G',T') \to (G'',T'')$, where $h' \longleftrightarrow \{h'^*, \eta'_{\alpha'}(\alpha' \in \mathcal{V}')\}$, then $h' \circ h \longleftrightarrow \{h'^* \circ h^*, \eta'_{h^*(\alpha)}\eta_\alpha\}$.

Proof: (i) Apply σ to both sides of equation (15); then (13) implies $h^\sigma(x_\alpha{}^\sigma(\xi)) = x'^\sigma_{h^*(\alpha)}(\eta_\alpha{}^\sigma\xi) = x'_{h^*\sigma(\alpha^\sigma)}(\xi'_{h^*(\alpha),\sigma}\,\eta_\alpha{}^\sigma\,\xi)$. Also by (13) and (15), one has

$$h^\sigma(x_\alpha{}^\sigma(\xi)) = h^\sigma(x_{\alpha^\sigma}(\xi_{\alpha,\sigma}\,\xi)) = x'_{h^{\sigma*}(\alpha^\sigma)}(\xi_{\alpha,\sigma}\,\mu_{\alpha^\sigma}\,\xi), \text{ where}$$

$h^\sigma \longleftrightarrow \{h^{\sigma*}, \mu_\alpha(\alpha \in \mathcal{V})\}$. Equating the two lines, it follows that $h^{*\sigma} = h^{\sigma*}$, and $\mu_{\alpha^\sigma} = \eta_\alpha{}^\sigma \dfrac{\xi'_{h^*(\alpha),\sigma}}{\xi_{\alpha,\sigma}}$. Replacing α by $\alpha^{\sigma^{-1}}$ in the argument, the assertion follows.

(ii) is verified in a similar way.

Corollary: The isomorphism h is defined over k if and only if $h^{*\sigma} = h^*$ and $\xi'_{h^*(\alpha),\sigma} = (\eta_\alpha{}^\sigma/\eta_\alpha{}^\sigma)\xi_{\alpha,\sigma}$ for all $\sigma \in \Gamma$ and $\alpha \in \nabla$.

We can apply our discussion above to the case where $G' = \underline{G} = G(X,\nabla)$ is a Chevalley group, and (G,f) is a K/k-form of $\underline{G}$ (see §2.1). Thus, identify X and $\underline{X}$ by $\chi = f^*(\chi)$ for all $\chi \in X$, and for each $\alpha \in \underline{\nabla}$ fix $\underline{x}_\alpha$ defined over k. If we define $x_\alpha = f^{-1} \circ \underline{x}_\alpha$ for all $\alpha \in \nabla$, then (replacing h by f and $x'_{\alpha'}$ by $\underline{x}_\alpha$ above) we see that

$$f \longleftrightarrow \{f^*,1\}, \qquad \text{(follows from (15))}$$

$$f^{-1} \longleftrightarrow \{f^{*-1},1\},$$

$$f^\sigma \longleftrightarrow \{f^{*\sigma},\xi^{-1}_{\alpha^{\sigma^{-1}},\sigma}\} \quad \text{(follows from Lemma 1,(i),}$$

and the fact that Γ operates trivially on $\underline{x}_\alpha$ implies $\underline{\xi}_{\alpha,\sigma} = 1$ for all $\sigma \in \Gamma$, (13)). Since $\varphi_\sigma = f^\sigma \circ f^{-1}$, Lemma 1,(ii) implies

$$\varphi_\sigma \longleftrightarrow \{\varphi_\sigma{}^*, \xi^{-1}_{\alpha^{\sigma^{-1}},\sigma}\}.$$

Since f is uniquely determined by the system $\{\varphi_\sigma\}$, it follows that the k-isomorphism classes of G are determined by the systems $\{\varphi_\sigma{}^*,\xi^{-1}_{\alpha^{\sigma^{-1}},\sigma}\}$, (I, Coroll. 3.1.4).

Lemma 2: With notations as above, the isomorphism $h: G \to G'$ is uniquely determined by the system $\{h^*,\eta_\alpha \ (\alpha \in \Delta)\}$, where Δ is a fundamental system of ∇.

We omit the proof of this lemma; by Lemma 1,(ii) the proof is reduced to the case where h is an automorphism of G, and this is discussed in [2],17-08,09. In the case of char. k = 0, a direct proof follows from the normalization of the basis of the Lie algebra of G; in the case of char. k = p one can use the normalization of Chevalley ("Sur Certains Groupes Simples," Tohoku Math. Jour., 1955). We indicate briefly how the η_α arise in the char. k = 0 case. First, one can normalize the x_α so that $x_\alpha(\xi) = \exp(\xi E_\alpha)$ where $\{E_\alpha\}$ is a 'Weyl basis' for $\mathfrak{g}$, the Lie algebra of G (one has $\mathfrak{g} = \mathfrak{t} + \Sigma\, \mathfrak{g}_\alpha$, where $\mathfrak{t}$ is the Lie algebra of T, and $\mathfrak{g}_\alpha$ are one-dimensional root-spaces). The basis $\{E_\alpha\}$ satisfies relations $[E_\alpha, E_\beta] = N_{\alpha,\beta} E_{\alpha+\beta}$ where α,β and $\alpha+\beta \in \mathcal{V}$, and $N_{\alpha,\beta} = -N_{-\alpha,-\beta} \in \mathbb{Z}$, $[E_\alpha, E_{-\alpha}] = H_\alpha$ where $H_\alpha \in \mathfrak{t}$ and satisfies $(H_\alpha, H) = \alpha^*(H)$ for all $H \in \mathfrak{t}$, with (,) the Killing form on $\mathfrak{t}$. The isomorphism $h:G \rightarrow G'$ induces an isomorphism of $\mathfrak{g} \rightarrow \mathfrak{g}'$, where $\mathfrak{g}'$ is the Lie algebra of G'; denoting this induced isomorphism as h also, one has $h(E_\alpha) = \eta_\alpha E'_{h^*(\alpha)}$, where $\{E'_{\alpha'}\}$ is a Weyl basis for $\mathfrak{g}'$. From this, one can obtain the relations

$$\eta_{\alpha+\beta} = \frac{N'_{h^*(\alpha),h^*(\beta)}}{N_{\alpha,\beta}} \eta_\alpha \eta_\beta,$$

$$\eta_{-\alpha} = \eta_\alpha^{-1}.$$

From these relations, it is clear that the η_α, $\alpha \in \mathcal{V}$ are determined by the η_{α_i}, $\alpha_i \in \Delta$.

<u>Lemma 3</u>: Let $\varphi \in \mathrm{Aut}(G,T)$. One has $\varphi \longleftrightarrow \{1, \eta_\alpha\}$ if and only if $\varphi = I_t$ for some $t \in T$, and $\eta_\alpha = \alpha(t)$.

<u>Proof</u>: It is clear from equations (15) and (1) that for any $t \in T$, $I_t \longleftrightarrow \{1, \alpha(t)\}$.

Conversely, suppose $\varphi \in \mathrm{Aut}(G,T)$, $\varphi \longleftrightarrow \{1, \eta_\alpha\}$. Since Δ is a basis for X, there exists some $t \in T$ satisfying $\alpha_i(t) = \eta_{\alpha_i}$ for all $\alpha_i \in \Delta$ (for the mapping $t \rightarrow (\alpha_1(t),\ldots,\alpha_\ell(t))$ is an isomorphism of T/Z onto $(\mathbb{G}_m)^\ell$, where Z = center G). Thus by Lemma 2, and the fact that $I_t \longleftrightarrow \{1, \alpha(t)\}$, it follows that $\varphi = I_t$ and $\eta_\alpha = \alpha(t)$ for all $\alpha \in \mathcal{V}$.

<u>Remark</u>: We can now complete the proof of Prop. 1.4.1. Recall that we had defined $\Theta' = \{\theta \in \mathrm{Aut}(G,T) | \theta^*(\Delta) = \Delta\}$, and we had shown that $\Theta' \cap \mathrm{Inn}(G) = T/Z$, and $\mathrm{Aut}(G) = \Theta' \, \mathrm{Inn}(G)$. Define $\Theta = \{\theta \in \mathrm{Aut}(G,T) | \theta \longleftrightarrow \{\theta^*, 1\}, \theta^*(\Delta) = \Delta\}$. Clearly $\Theta \subset \Theta'$, and by Lemma 3, $\Theta \cap \mathrm{Inn}(G) = \{1\}$. Also, Lemma 1,(ii) implies that if $\theta' \in \Theta'$, where $\theta' \leftrightarrow \{\theta'^*, \eta_\alpha\}$, then $\theta' = \theta \circ I_t$ with $\theta \in \Theta$, where $\theta \leftrightarrow \{\theta'^*, 1\}$, and $I_t \longleftrightarrow \{1, \eta_\alpha\}$. Thus $\mathrm{Aut}(G) = \Theta \, \mathrm{Inn}(G)$, and this product is semi-direct. It should also be noted that if T is k-trivial, then all $\theta \in \Theta$ are defined over k.

3° <u>Definition</u>: Let G, G', T, T', etc. be as in the theorem. If ρ is an isomorphism of $(X, \surd)$ onto $(X', \surd')$ and $\{\eta_\alpha\}_{\alpha \in \surd}$ a set of scalars in K, then the system $\{\rho, \eta_\alpha \; (\alpha \in \surd)\}$ will be called <u>admissible</u> if there is a K-isomorphism $h:(G,T) \rightarrow (G',T')$ such that $\rho = h^*$ and $h \leftrightarrow \{h^*, \eta_\alpha(\alpha \in \surd)\}$.

Note that Lemma 2 may be restated as: If the system $\{h^*, \eta_\alpha \; (\alpha \in \surd)\}$ is admissible, then the η_α are uniquely determined by $\{\eta_{\alpha_i} \; (\alpha_i \in \Delta)\}$.

<u>Lemma 4</u>: If ρ is any isomorphism of $(X, \surd)$ onto $(X', \surd')$ and $\{\eta_{\alpha_i} \; (\alpha_i \in \Delta)\}$ any set of scalars in K*, then there exists $\{\eta_\alpha'(\alpha \in \surd)\}$, a unique set of scalars in K* satisfying $\eta'_{\alpha_i} = \eta_{\alpha_i}$ for all $\alpha_i \in \Delta$, and such that the system $\{\rho, \eta_\alpha'(\alpha \in \surd)\}$ is admissible.

<u>Proof:</u> By Theorem 1.3.1, there is an isomorphism $h_1 : G \rightarrow G'$, defined over some finite extension K_1 of k, such that $h_1^* = \rho$. Let $h_1 \longleftrightarrow \{\rho, \mu_\alpha \; (\alpha \in \surd)\}$; then as in the proof of Lemma 3, there is an element $t \in T$ such that $\alpha_i(t) = \mu_{\alpha_i}^{-1} \eta_{\alpha_i}$ for all $\alpha_i \in \Delta$. Define $h = h_1 \circ I_t$ and $\eta_\alpha' = \alpha(t)\mu_\alpha$; then Lemma 1,(ii) and Lemma 2 imply $h \longleftrightarrow \{\rho, \eta_\alpha'\}$, so the system $\{\rho, \eta_\alpha'\}$ is admissible. Since t is K_2-rational, where K_2 is a

finite extension of k, and α is defined over K, it follows that the $\eta_\alpha' \in K'$, where K' is a finite extension of k containing K, K_1, K_2. We show later (Cor. 2.4.2) that one can take K' = K.

4° We now prove Theorem 2.4.1.

Let $h_o \leftrightarrow \{h_o^*, \eta_\alpha^o\ (\alpha \in \mathcal{V}_o)\}$. It suffices to show there exists an admissible system $\{h^*, \eta_\alpha\ (\alpha \in \mathcal{V})\}$ satisfying

(a) $\begin{cases} h^{*\sigma} = h^* & \text{all } \sigma \in \Gamma \\ \xi'_{h^*(\alpha),\sigma} = \dfrac{\eta_\alpha{}^\sigma}{\eta_\alpha{}^\sigma}\,\xi_{\alpha,\sigma} & \text{for all } \sigma \in \Gamma,\ \alpha \in \mathcal{V} \end{cases}$ (see Coroll., p.90),

(b) $h^*|X^o = h_o^*$, $\eta_\alpha = \eta_\alpha^o$ for all $\alpha \in \mathcal{V}_o$,

(c) There exists $w' \in W'^\Gamma$ such that $\rho = w'h^*$ (thus $h^{[*]} = \rho$).

First, since ρ_o is associated with h_o, there exists an element $w' \in W_o'$ such that $\rho_o = w'h_o^* = h_o^{[*]}$. Define $h^* = w'^{-1}\rho$. It is clear that h^* is an isomorphism of $(X,\mathcal{V})$ onto $(X',\mathcal{V}')$ satisfying $h^*|X^o = h_o^*$.

Next, we show that $h^{*\sigma} = h^*$ for all $\sigma \in \Gamma$. Since $X^o_{\mathbb{Q}}$ can be identified with $\{\mathcal{V}_o\}_{\mathbb{Q}}$ as a vector subspace of $X_{\mathbb{Q}}$ and $X_{\mathbb{Q}} = X^o_{\mathbb{Q}} + X^{o\perp}_{\mathbb{Q}}$, it is enough to show that $h^*|\{\mathcal{V}_o\}_{\mathbb{Q}}$ and $h^*|\{\mathcal{V}_o\}^\perp_{\mathbb{Q}}$ are Γ-invariant. Since $h^*|\{\mathcal{V}_o\}_{\mathbb{Q}} = h_o^*$, and by hypothesis, h_o is defined over k, Lemma 1 implies $h_o^{*\sigma} = h_o^*$ for all $\sigma \in \Gamma$. Now $\{\mathcal{V}_o\}^\perp_{\mathbb{Q}}$ and $\{\mathcal{V}_o'\}^\perp_{\mathbb{Q}}$ can be identified with $Y_{\mathbb{Q}}$ and $Y'_{\mathbb{Q}}$, and Γ acts trivially on Y and Y', so that to show $h^*|\{\mathcal{V}_o\}^\perp_{\mathbb{Q}}$ is Γ-invariant, we only need to show $h^* \circ \sigma = \sigma \circ h^*$ on $\{\mathcal{V}_o\}^\perp_{\mathbb{Q}}$, for all $\sigma \in \Gamma$. By definition, $[\sigma] = w_\sigma^{-1} \circ \sigma$, $[\sigma]' = w_\sigma'^{-1} \circ \sigma$, with $w_\sigma \in W_o$, $w_\sigma' \in W_o'$, and $h^* = w'^{-1}\rho$ with $w' \in W_o'$; but W_o and W_o' act trivially on Y and Y' respectively, so $[\sigma]|Y_{\mathbb{Q}} = \sigma$, $[\sigma]'|Y'_{\mathbb{Q}} = \sigma$, $h^*|Y = \rho$. Since ρ is a congruence, $[\sigma]' = \rho[\sigma]\rho^{-1}$, hence $\sigma = h^* \circ \sigma \circ h^{*-1}$ on $Y'_{\mathbb{Q}}$, which proves the assertion.

Finally, define $\eta_{\alpha_i} = \eta^{o}_{\alpha_i}$ for all $\alpha_i \in \Delta_o$. By Lemma 4, if η_{α_i} for $\alpha_i \in \Delta - \Delta_o$ are arbitrary constants in K*, then the set of scalars $\{\eta_{\alpha_i}\}_{\alpha_i \in \Delta}$ can be uniquely extended so that $\{h^*, \eta_\alpha \ (\alpha \in \mathcal{V})\}$ is an admissible system. Since any such system satisfies (b), we only need to show that we can choose the η_{α_i} for $\alpha_i \in \Delta - \Delta_o$ so that the second condition in (a) is satisfied; this will complete the proof of the theorem.

For the moment, let η_{α_i}, $\alpha_i \in \Delta - \Delta_o$ be arbitrary scalars in K*, and $\{h^*, \eta_\alpha \ (\alpha \in \mathcal{V})\}$ the admissible system determined by them. Define

$$(16) \qquad \zeta_{\alpha,\sigma} = \frac{\eta^{\sigma}_{\alpha}}{\eta_{\alpha^\sigma}} \frac{\xi'_{h^*(\alpha),\sigma}}{\xi_{\alpha,\sigma}}.$$

We must show that the η_{α_i}, $\alpha_i \in \Delta - \Delta_o$ can be modified so that $\zeta_{\alpha,\sigma} = 1$ for all $\alpha \in \mathcal{V}$, $\sigma \in \Gamma$; of course, by Coroll. (p.90), this is already true for all $\alpha \in \mathcal{V}_o$, $\sigma \in \Gamma$. Let $h \longleftrightarrow \{h^*, \eta_\alpha \ (\alpha \in \mathcal{V})\}$. Since $h^{*\sigma} = h^*$, all $\sigma \in \Gamma$, equation (16) and Lemma 1 imply $h^\sigma \longleftrightarrow \{h^*, \zeta_{\alpha^{\sigma^{-1}},\sigma}\, \eta_\alpha\}$ and $h^\sigma \circ h^{-1} \longleftrightarrow \{1, \zeta_{\alpha^{\sigma^{-1}},\sigma}\}$. Thus by Lemma 3, $h^\sigma \circ h^{-1} = I_{t'_\sigma}$ for some $t'_\sigma \in T'$ and $\zeta_{\alpha^{\sigma^{-1}},\sigma} = \alpha(t'_\sigma)$. From this last equation, one sees easily that

$$(17) \qquad \begin{cases} \zeta_{-\alpha,\sigma} = \zeta^{-1}_{\alpha,\sigma} \\ \zeta_{\alpha+\beta,\sigma} = \zeta_{\alpha,\sigma}\zeta_{\beta,\sigma}. \end{cases}$$

From (17), it is clear that if we can choose η_{α_i}, $\alpha_i \in \Delta - \Delta_o$ so that $\zeta_{\alpha_i,\sigma} = 1$ for all $\sigma \in \Gamma$, then $\zeta_{\alpha,\sigma} = 1$ for all $\alpha \in \mathcal{V}$, $\sigma \in \Gamma$. Also, (17) (or rather the relation $\zeta_{\alpha^{\sigma^{-1}},\sigma} = \alpha(t'_\sigma)$) shows that $\zeta_{\alpha,\sigma}$ depends only on $\alpha \bmod (\mathcal{V}_o)_{\mathbb{Z}}$. From equations (16) and (14), one can easily verify the following condition:

$$(18) \qquad \zeta^{\tau}_{\alpha,\sigma}\, \zeta_{\alpha^\sigma,\tau} = \zeta_{\alpha,\sigma\tau}.$$

Now fix $\alpha_i \in \Delta - \Delta_o$, and define $\bar{\Gamma}_{\alpha_i} = \{\sigma \in \Gamma | \alpha_i^{[\sigma]} = \alpha_i\}$. We first show that η_{α_i} can be modified so that $\zeta_{\alpha_i,\sigma} = 1$ for all $\sigma \in \bar{\Gamma}_{\alpha_i}$. If $\sigma \in \bar{\Gamma}_{\alpha_i}$, then $\alpha_i = \alpha_i^{[\sigma]} = w_\sigma^{-1}\alpha_i^\sigma = \alpha_i^\sigma + \chi_o$, with $\chi_o \in \{\mathcal{V}_o\}_{\mathbb{Z}}$ since $w_\sigma \in W_o$. But then $\zeta_{\alpha_i^\sigma,\tau} = \zeta_{\alpha_i,\tau}$, and so by (18), the system $(\zeta_{\alpha_i,\sigma})$ is a one-cocycle of $\bar{\Gamma}_{\alpha_i}$ in K*. By Hilbert's Theorem 90, there is an element $\mu \in K^*$ satisfying $\zeta_{\alpha_i,\sigma} = \mu^\sigma\mu^{-1}$; if η_{α_i} is replaced by $\mu^{-1}\eta_{\alpha_i}$, then (16) implies $\zeta_{\alpha_i,\sigma} = 1$ for all $\sigma \in \bar{\Gamma}_{\alpha_i}$. Let $\{\sigma_1,\ldots,\sigma_m\}$ be a complete set of coset representatives of $\bar{\Gamma}_{\alpha_i}$ in Γ, with $\sigma_1 = 1$; then $\{\alpha_i^{[\sigma_j]}\}_{1\le j\le m}$ is the orbit of α_i by the [ ] operation of Γ. Next we show that for each $j \neq 1$, one can adjust $\eta_{\alpha_i^{[\sigma_j]}}$ so that $\zeta_{\alpha_i,\sigma_j} = 1$ for $2 \le j \le m$. If $j \neq 1$, then $\alpha_i^{[\sigma_j]} \in \Delta$ and $\alpha_i^{[\sigma_j]} \neq \alpha_i$. By definition, the root $\beta = w_{\sigma_j}^{-\sigma_j^{-1}}\alpha_i$ satisfies $\beta^{\sigma_j} = \alpha_i^{[\sigma_j]}$, so that by (16),

$$\zeta_{\beta,\sigma_j} = \frac{\eta_\beta^{\sigma_j}}{\eta_{\alpha_i^{[\sigma_j]}}}\frac{\xi'_{h^*(\beta),\sigma_j}}{\xi_{\beta,\sigma_j}}.$$

Now replace $\eta_{\alpha_i^{[\sigma_j]}}$ by a scalar so that $\zeta_{\beta,\sigma_j} = 1$. Since $\beta = \alpha_i \bmod (\mathcal{V}_o)_{\mathbb{Z}}$, it follows that $\zeta_{\alpha_i,\sigma_j} = 1$. Finally, we show that these modifications in the scalars $\eta_{\alpha_i^{[\sigma_j]}}$, $1 \le j \le m$ imply $\zeta_{\alpha_i^{[\sigma_j]},\tau} = 1$ for all $\tau \in \Gamma$, $1 \le j \le m$. If $\tau \in \Gamma$, then $\tau = \sigma\sigma_j$ for some j and some $\sigma \in \bar{\Gamma}_{\alpha_i}$, so by (18), one has

$$\zeta_{\alpha_i,\tau} = \zeta_{\alpha_i,\sigma\sigma_j} = \zeta_{\alpha_i,\sigma}^{\sigma_j}\,\zeta_{\alpha_i^\sigma,\sigma_j} = 1, \text{ all } \tau \in \Gamma,$$

(since $\zeta_{\alpha_i^\sigma,\sigma_j} = \zeta_{\alpha_i,\sigma_j} = 1$). But then (18) implies $\zeta_{\alpha_i^{\sigma_j},\tau} = 1$ for all $\tau \in \Gamma$, all j, and so

$$\zeta_{\alpha_i^{[\sigma_j]},\tau} = \zeta_{w_{\sigma_j}^{-1}\alpha_i^{\sigma_j},\tau} = \zeta_{\alpha_i^{\sigma_j},\tau} = 1.$$

Thus to choose the η_{α_i}, $\alpha_i \in \Delta - \Delta_0$, one decomposes $\Delta - \Delta_0$ into orbits under the [ ]-operation of Γ, and in each orbit, fixes a representative α_i. Using the method described above, one chooses the scalars $\eta_{\alpha_i[\sigma j]}$ where $\alpha_i^{[\sigma j]}$ runs through all elements in the orbit of α_i. Since these choices give $\zeta_{\alpha_K,\tau} = 1$ for all $\alpha_K \in \Delta - \Delta_0$, $\tau \in \Gamma$, the proof of the theorem is completed.

In order to describe some immediate consequences of Theorem 2.4.1, we need a definition.

<u>Definition</u>: A connected semi-simple algebraic group G defined over k is said to be of <u>Chevalley type over</u> k (or k-split) if G contains a maximal torus T which is k-trivial (i.e., A = T).

A connected semi-simple algebraic group G defined over k is said to be of <u>Steinberg type over</u> k (or k-quasi-split) if G contains a Borel subgroup which is defined over k.

<u>Remark 1</u>: If G is of Chevalley type over k, then G is of Steinberg type over k. In fact, if T is a maximal torus of G which is k-trivial, then every Borel subgroup of G containing T is defined over k (since $B_\Delta^\sigma = B_{\Delta^\sigma} = B_\Delta$ for all $\sigma \in \Gamma$). Conversely if every Borel subgroup of G containing a torus T/k is defined over k, then T is k-trivial, and so G is of Chevalley type over k. (See Ono, "On the field of definition of Borel subgroups of semi-simple algebraic groups," Jour. Math. Soc. Japan, Vol. 15, no. 4, 1963).

<u>Remark 2</u>: The following three conditions are equivalent:

(i) G is of Steinberg type over k ,

(ii) $Z(A) = T$,

(iii) $\mathcal{V}_0 = \emptyset$.

The equivalence of (i) and (ii) follows from the two facts: (a) Z(A) is the reductive part of a minimal parabolic subgroup of G defined over k and containing T, and (b) T is the reductive part of a Borel subgroup of G. The equivalence of (ii) and (iii) follows from the fact that the semi-simple part of Z(A) is generated by P_α, $\alpha \in \sqrt{}_o$.

Note that (iii) can occur even though $X_o \neq \{0\}$, that is, G is of Steinberg type but not of Chevalley type if $\sqrt{}_o = \emptyset$ and the action of Γ on Δ is non-trivial (recall X_o is generated by $\alpha_i^\sigma - \alpha_i$, $\alpha_i \in \Delta$, $\sigma \in \Gamma$).

If G is of Steinberg type over k, then (iii) above implies $\Delta_o = \emptyset$, and hence $G_o = \{1\}$. Thus the Γ-diagram of G is of the form $(X,\Delta,[\sigma])$, where $[\sigma]$ is just the ordinary action of Γ on Δ (since $W_o = \{1\}$). If G is of Chevalley type over k, then in addition, Γ acts trivially on Δ, so that the Γ-diagram of G is just the ordinary Dynkin diagram (X,Δ). Thus we have the following corollary to theorem 2.4.1:

<u>Corollary 2.4.2</u>:

(1) The k-isomorphism class of a group of Chevalley type over k is uniquely determined by (X,Δ). (In particular, every such group is k-isomorphic to a Chevalley group.)

(2) The k-isomorphism class of a group of Steinberg type over k is uniquely determined by its Γ-diagram $(X,\Delta,[\sigma])$.

Since the complete classification of Chevalley groups is known, (1) above states the classification problem over k for groups of Chevalley type over k is solved. Part (2) of the corollary reduces the classification problem for groups of Steinberg type over k to a problem of classifying Γ-diagrams with $\Delta_o = \emptyset$. Later we will show that if $k = \mathbb{F}_q$, a finite field, then <u>any</u> connected semi-simple algebraic group defined over k is of Steinberg type.

§3. Classification of semi-simple algebraic groups defined over k (k perfect)

Our notations remain as in §2 (unless specified otherwise).

3.1 Investigation of Γ-diagrams

Definition: If X is a free module of rank ℓ, Δ a fundamental system of a root system $\mathcal{V}$ in X, Δ_0 a subset of Δ, and [ ] a homomorphism of the Galois group Γ into $\mathrm{Aut}(X,\Delta,\Delta_0)$, we will say that the system $\mathcal{S} = (X,\Delta,\Delta_0,[\sigma])$ is admissible if there exists a connected semi-simple group G defined over k having $\mathcal{S}$ as Γ-diagram.

We are only interested, of course, in classifying admissible systems $\mathcal{S}$. If $\mathcal{S}$ is a system $(X,\Delta,\Delta_0,[\sigma])$, then one obtains a subsystem $\mathcal{S}_0 = \{X^0,\Delta_0,[\sigma]\}$, where X^0 is the projection of X on $\{\Delta_0\}_{\mathbb{Q}}$ (one may write $X_{\mathbb{Q}} = \{\Delta_0\}_{\mathbb{Q}} + \{\Delta_0\}_{\mathbb{Q}}^{\perp}$ with respect to some W-invariant metric), and $\mathcal{S}_0$ is just the Γ-diagram of (G_0,T^0), the k-compact kernel of the group (G,T) having $\mathcal{S}$ as Γ-diagram. For this section, we will only be interested in systems $\mathcal{S}$ for which the subsystem $\mathcal{S}_0$ is admissible and corresponds to a k-compact group (G_0,T^0) where T^0 splits over K (we will write $(G_0,T^0) \to \mathcal{S}_0$). (In order to investigate the general problem of classifying Γ-diagrams, we are assuming for the moment that the special case of classifying admissible diagrams of k-compact groups is solved.)

Definition: The system $\mathcal{S}$ is said to be admissible over $(G_0,T^0) \to \mathcal{S}_0$ if there is a connected semi-simple algebraic group G defined over k and a maximal torus T defined over k such that (G_0,T^0) is the k-compact kernel of (G,T), and $\mathcal{S}$ is the Γ-diagram of G.

Theorem 2.4.1 states that if $\mathcal{S}$ is admissible over $(G_0,T^0) \to \mathcal{S}_0$, then the group (G,T) as described in the above definition is unique up to k-isomorphism.

We wish to obtain conditions under which a system $\mathcal{S}$ is admissible over $(G_o,T^o) \to \mathcal{S}_o$. Let $\mathcal{S} = \{X,\Delta,\Delta_o,[\sigma]\}$ and suppose $\mathcal{S}$ is admissible over $(G_o,T^o) \to \mathcal{S}_o$. The Dynkin diagram (X,Δ) uniquely determines (up to k-isomorphism) a Chevalley group $(\underline{G},\underline{T})$ defined over k. $\mathcal{S}$ is the Γ-diagram of (G,T), a connected semi-simple group defined over k having (G_o,T^o) as k-compact kernel, and (G,T) is a K/k-form of $(\underline{G},\underline{T})$. The K-isomorphism $f:(G,T) \to (\underline{G},\underline{T})$ is uniquely determined by the system $\{\varphi_\sigma\}$ $(\varphi_\sigma = f^\sigma \circ f^{-1} \in \mathrm{Aut}_K(\underline{G},\underline{T}))$, and we have seen in §2.4, 2°, that $\varphi_\sigma \leftrightarrow \{\varphi_\sigma{}^*, \xi^{-1}_{\alpha^{\sigma^{-1}},\sigma}\}$. The cocycle condition: $\varphi_\sigma{}^\tau \circ \varphi_\tau = \varphi_{\sigma\tau}$ implies that $\varphi_\sigma{}^*\varphi_\tau{}^* = \varphi^*_{\sigma\tau}$ and (14). Since we are assuming $(G_o,T^o) \to \mathcal{S}_o$, the scalars $\{\xi^{-1}_{\alpha^{\sigma^{-1}},\sigma},\ \alpha \in \mathcal{V}_o,\ \sigma \in \Gamma\}$ are given, and so is $\varphi_\sigma{}^*|X^o$. Since $\mathcal{S}$ is admissible, Theorem 2.4.1 (applied to the case $G' = \underline{G}$) implies that the set of scalars $\{\xi^{-1}_{\alpha^{\sigma^{-1}},\sigma},\ \alpha \in \mathcal{V},\sigma \in \Gamma\}$ is determined by the subset $\{\xi^{-1}_{\alpha^{\sigma^{-1}},\sigma},\ \alpha \in \mathcal{V}_o,\ \sigma \in \Gamma\}$. Also, the isomorphisms φ^*_σ are determined by the Γ-diagram $\mathcal{S}$ and the restrictions $\varphi_\sigma{}^*|X^o$. For, under the identification of X with $\underline{X}$ (via f*) we have $\chi^{[\sigma]} = w_\sigma^{-1}\chi^\sigma = w_\sigma^{-1}\varphi_\sigma{}^{*-1}(\chi)$ so $\varphi_\sigma{}^* = [\sigma]^{-1}w_\sigma{}^{-1}$. But $[\sigma]$ is given by $\mathcal{S}$, and since $w_\sigma \in W_o$, $\varphi_\sigma{}^*|X^o$ determines w_σ. From these observations, we see that a <u>necessary</u> condition for $\mathcal{S}$ to be admissible over $(G_o,T^o) \to \mathcal{S}_o$ is that the set of scalars $\{\xi^{-1}_{\alpha^{\sigma^{-1}},\sigma},\ \alpha \in \mathcal{V}_o,\ \sigma \in \Gamma\}$ can be extended to a set of scalars $\{\xi^{-1}_{\alpha^{\sigma^{-1}},\sigma},\ \alpha \in \mathcal{V},\ \sigma \in \Gamma\}$ which satisfies (14) and also satisfies the condition that, for each σ, if one defines $\varphi_\sigma{}^* = [\sigma]^{-1}w_\sigma{}^{-1}$, then $\{\varphi^*_\sigma,\xi^{-1}_{\alpha^{\sigma^{-1}},\sigma}\}$ is admissible in the sense of §2.4, 3°. This condition is also <u>sufficient</u>. For, if $\{\varphi_\sigma{}^*, \xi^{-1}_{\alpha^{\sigma^{-1}},\sigma}\}$ is admissible for each σ, then a system $\{\varphi_\sigma\}$ of automorphisms of $(\underline{G},\underline{T})$ is determined, and by equation (14) and the definition of $\varphi_\sigma{}^*$, it may be verified that the system $\{\varphi_\sigma\}$ is a one-cocycle of Γ in $\mathrm{Aut}_K(\underline{G},\underline{T})$, hence $\{\varphi_\sigma\}$ determines a K/k-form of $(\underline{G},\underline{T})$. Thus $\mathcal{S}$ is admissible over $(G_o,T^o) \to \mathcal{S}_o$.

In chapter I, §4.4, we have defined the k-rank of G; we see by Prop. 2.1.6 that the k-rank of G is just the number of restricted fundamental roots of $\mathcal{V}$ (i.e., $\#\bar{\Delta}$). We wish to reduce our problem of classifying admissible Γ-diagrams to the case of Γ-diagrams of groups having k-rank = 1.

Let $\mathcal{S} = \{X,\Delta,\Delta_o,[\sigma]\}$. If Δ' is a $[\sigma]$-invariant subset of Δ, then we can define a subsystem $\mathcal{S}_{\Delta'} = \{X',\Delta',\Delta_o',[\sigma]'\}$ of $\mathcal{S}$ where X' is the projection of X on $\{\Delta'\}_{\mathbb{Q}}$, $\Delta_o' = \Delta' \cap \Delta_o$, $[\sigma]' = [\sigma]|X'$. Such a system $\mathcal{S}_{\Delta'}$ will be called a <u>canonical</u> <u>subsystem</u> of $\mathcal{S}$. If $\mathcal{S}$ is admissible, and is the Γ-diagram of (G,T), then Δ' determines a connected semi-simple subgroup $G(\Delta')$ of G, namely, the subgroup generated by $\{P_\alpha | \alpha \in \mathcal{V} \cap \{\Delta'\}_{\mathbb{Q}}\}$.

<u>Lemma 3.1.1</u>: The subgroup $G(\Delta')$ is defined over k if the following two conditions are satisfied:

(i) $\Delta'^{[\sigma]} = \Delta'$,

(ii) if $\alpha \in \Delta'$ and $\beta \in \Delta_o$ and $\langle\alpha,\beta\rangle \neq 0$, then $\beta \in \Delta'$.

<u>Proof</u>: From the definition of $G(\Delta')$, it is clear that $G(\Delta')$ is defined over k if $\mathcal{V}'^\sigma = \mathcal{V}'$, hence it suffices to show that if $\alpha_i \in \Delta'$, then $\alpha_i{}^\sigma \in \mathcal{V}'$ for all $\sigma \in \Gamma$. Now $\alpha_i{}^\sigma = w_\sigma \alpha_i^{[\sigma]}$, where $w_\sigma = w_{\alpha_{i_s}} \cdots w_{\alpha_{i_1}}$, $\alpha_{i_j} \in \Delta_o$, so define $\alpha'_K = w_{\alpha_{i_K}} \cdots w_{\alpha_{i_1}} \alpha_i^{[\sigma]}$, $1 \le K \le s$, and $\alpha_o' = \alpha_i^{[\sigma]}$. We show by induction that $\alpha_K' \in \mathcal{V}'$. By condition (i), $\alpha_i^{[\sigma]} \in \Delta'$, so we may assume that $\alpha'_{K-1} \in \mathcal{V}'$. If $\langle\alpha_{i_K},\Delta'\rangle = 0$, then $\alpha'_K = \alpha'_{K-1}$ (for, $\alpha'_K = w_{\alpha_{i_K}}\alpha'_{K-1}$, and $w_{\alpha_{i_K}}\alpha'_{K-1} = \alpha'_{K-1} - 2\dfrac{\langle\alpha_{i_K},\alpha'_{K-1}\rangle}{\langle\alpha_{i_K},\alpha_{i_K}\rangle}\alpha_{i_K}$). If

$\langle \alpha_{i_K}, \Delta' \rangle \neq 0$, then by condition (ii), $\alpha_{i_K} \in \Delta'$, which implies $\alpha'_K \in \mathcal{V}'$.

If $\mathcal{S}$ is admissible, and $(G,T) \to \mathcal{S}$, and Δ' is a subset of Δ which satisfies conditions (i),(ii) of the lemma, then the canonical subsystem $\mathcal{S}_{\Delta'}$ is clearly admissible, and $(G(\Delta'),T') \to \mathcal{S}_{\Delta'}$, where $T' = T \cap G(\Delta')$ is a maximal torus of $G(\Delta')$ which contains A', a maximal k-trivial torus in $G(\Delta')$. ($A' = A \cap T'$ is a maximal k-trivial torus in $G(\Delta')$, because the centralizer of A' in $G(\Delta')$ is equal to $T' \cdot G(\Delta_o')$ and $G(\Delta_o')$ is k-compact.)

<u>Proposition 3.1.2</u>: Let $\mathcal{S} = (X,\Delta,\Delta_o,[\sigma])$, and suppose $\Delta = \Delta' \cup \Delta''$, where Δ' and Δ'' satisfy conditions (i),(ii) of lemma 3.1.1 and $\Delta' \cap \Delta'' \subset \Delta_o$. If the canonical subsystems $\mathcal{S}_{\Delta'}$, $\mathcal{S}_{\Delta''}$ are admissible over $G_o(\Delta_o')$, $G_o(\Delta_o'')$, respectively, where $\Delta_o' = \Delta' \cap \Delta_o$, $\Delta_o'' = \Delta'' \cap \Delta_o$, then $\mathcal{S}$ is admissible over $(G_o,T^o) \to \mathcal{S}_o$.

<u>Remark</u>: Condition (ii) on Δ' and Δ'' implies that Δ_o' and Δ_o'' consist of unions of connected components of Δ_o, hence $G_o(\Delta_o')$ and $G_o(\Delta_o'')$ are normal subgroups of G_o ([2], exposé 17).

<u>Example</u>: Later we will show that the following Γ-diagrams are admissible over $\mathbb{R}$ or $\mathbb{Q}_p$:

Thus by Proposition 3.1.2, the following Γ-diagram for E_7 is admissible:

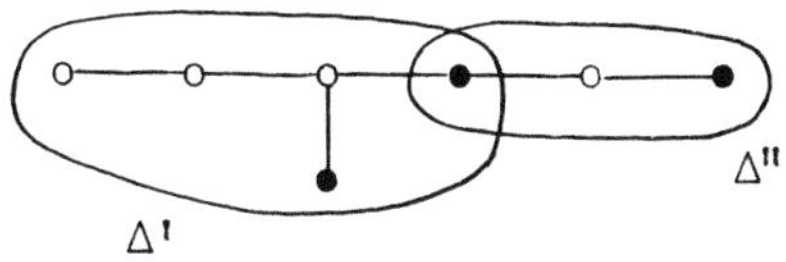

<u>Proof of Prop. 3.1.2</u>: We apply our remarks made at the beginning of this section. Let $\mathcal{V}' = \mathcal{V} \cap \{\Delta'\}_{\mathbb{Q}}$, $\mathcal{V}'' = \mathcal{V} \cap \{\Delta''\}_{\mathbb{Q}}$. Since we assume

$(G_o, T^o) \to \mathscr{S}_o$, and since $\mathscr{S}_{\Delta'}$ and $\mathscr{S}_{\Delta''}$ are admissible over $G_o(\Delta_o')$, $G_o(\Delta_o'')$, these determine three systems of scalars, namely $\{\xi_{\alpha,\sigma}\ (\alpha \in \mathscr{V}_o)\}$, $\{\xi'_{\alpha,\sigma}\ (\alpha \in \mathscr{V}')\}$, $\{\xi''_{\alpha,\sigma}\ (\alpha \in \mathscr{V}'')\}$, which satisfy (14) and which coincide for common α. For each $\alpha_i \in \Delta$ and $\sigma \in \Gamma$, define

$$\xi^{-1}_{\alpha_i^{\sigma^{-1}},\sigma} = \begin{cases} \xi'_{\alpha_i^{\sigma^{-1}},\sigma} & \text{if } \alpha_i \in \Delta' \\ \xi''_{\alpha_i^{\sigma^{-1}},\sigma} & \text{if } \alpha_i \in \Delta''. \end{cases}$$

For each $\sigma \in \Gamma$, the set of scalars $\{\xi^{-1}_{\alpha_i^{\sigma^{-1}},\sigma}\ (\alpha_i \in \Delta)\}$ can be extended so that the system $\{\varphi_\sigma^*, \xi^{-1}_{\alpha_i^{\sigma^{-1}},\sigma}\ (\alpha \in \mathscr{V})\}$ is admissible (Lemma 4, §2.4), and this extension does not conflict with the given $\{\xi_{\alpha,\sigma}\ (\alpha \in \mathscr{V}_o)\}$. Thus we only need to show that (14) is satisfied by the set of scalars $\{\xi^{-1}_{\alpha_i^{\sigma^{-1}},\sigma},\ \alpha \in \mathscr{V},\ \sigma \in \Gamma\}$ defined in this way. If we put $\zeta_{\alpha;\sigma,\tau} = \xi^{\tau}_{\alpha,\sigma}\, \xi_{\alpha^\sigma,\tau}\, \xi^{-1}_{\alpha,\sigma\tau}$, we only need to show $\zeta_{\alpha;\sigma,\tau} = 1$. Now $\varphi_\sigma \longleftrightarrow \{\varphi_\sigma^*, \xi^{-1}_{\alpha^{\sigma^{-1}},\sigma}\ (\alpha \in \mathscr{V})\}$ with $\varphi_\sigma \in \mathrm{Aut}(\underline{G}, \underline{T})$, and by Lemma 1, §2.4, one has $\varphi_\sigma^{\tau} \varphi_\tau \varphi^{-1}_{\sigma\tau} \longleftrightarrow \{1, \zeta^{-1}_{\alpha^{\tau^{-1}\sigma^{-1}};\sigma,\tau}\}$. Thus by Lemma 3, §2.4, $\zeta^{-1}_{\alpha^{\tau^{-1}\sigma^{-1}};\sigma,\tau} = \alpha(t_{\sigma,\tau})$ for some $t_{\sigma,\tau} \in \underline{T}$, and from this, we see that $\zeta_{\alpha;\sigma,\tau}$ is additive in α. Since $\zeta_{\alpha;\sigma,\tau} = 1$ if $\alpha_i \in \Delta' \cup \Delta'' = \Delta$, we see that $\zeta_{\alpha;\sigma,\tau} = 1$, for all $\alpha \in \mathscr{V}$.

By Proposition 3.1.2, the classification of admissible Γ-diagrams is reduced to the case of k-rank = 1.[11)]

Definition: The system $\mathscr{S} = (X, \Delta, \Delta_o, [\sigma])$ is k-*irreducible* if Δ is not the union of two mutually orthogonal $[\sigma]$-invariant (non-empty) subsystems Δ', Δ''. The system $\mathscr{S}$ is *absolutely irreducible* if Δ is connected. The system $\mathscr{S}$ is *simply connected* if $X = \{\Delta^*\}^\wedge_{\mathbb{Z}}$, and *adjoint* if $X = \{\Delta\}_{\mathbb{Z}}$.

We show that the classification of admissible Γ-diagrams can, in fact, be reduced to the case of absolutely irreducible diagrams of k-rank $= 1$. Suppose $\mathscr{S} = (X,\Delta,\Delta_o,[\sigma])$ is simply connected. If $\mathscr{S}$ is k-irreducible, but not absolutely irreducible, then $\Delta = \Delta_1 \cup \ldots \cup \Delta_s$, where the Δ_i are mutually disjoint connected components of Δ and correspondingly one has $X = X_1 + \ldots + X_s$. Define $\Gamma_1 = \{\sigma \in \Gamma \mid \Delta_1^{[\sigma]} = \Delta_1\}$; then $\Gamma = \bigcup_{i=1}^{s} \Gamma_1\sigma_i$, where $\Delta_i = \Delta_1^{[\sigma_i]}$. Let $\mathscr{S}_1 = (X_1,\Delta_1,\Delta_1 \cap \Delta_o,[\sigma])$, where $\sigma \in \Gamma_1$ and let k_1 be the fixed field of Γ_1. From our remarks at the end of section 1, we have that $\mathscr{S}$ is admissible as a Γ-diagram if and only if $\mathscr{S}_1$ is admissible as a Γ_1-diagram, and where $(G_1,T_1)/k_1 \to \mathscr{S}_1$, one has $(G,T) = R_{k_1/k}(G_1,T_1) \to \mathscr{S}$. Thus the classification is reduced to absolutely irreducible diagrams. By Prop. 3.1.2, we may further assume that k-rank $= 1$.[12]

Example: Steinberg groups

For groups of Steinberg type, the only absolutely irreducible k-rank $= 1$ diagrams consist of either a single vertex o, or . (We will show in the next section that if k is finite, then any connected semi-simple algebraic group defined over k is of Steinberg type.)

The only new diagrams, that is, the Γ-diagrams of Steinberg groups which are not Chevalley groups are those where the action of Γ on Δ is non-trivial. Thus the only possibilities for new Γ-diagrams are:

$^2A_\ell$: 

$^2D_\ell$: 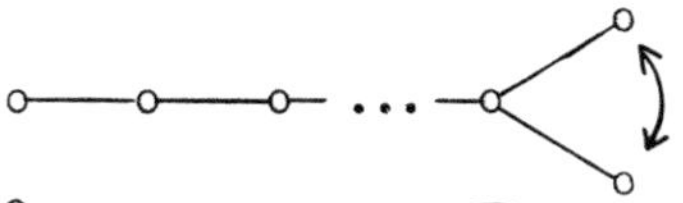

$^3D_4$: 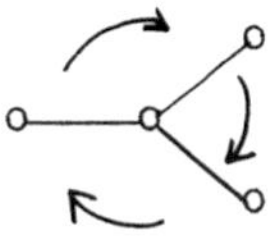

$^6D_4$: 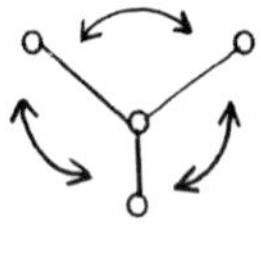 (i.e., the action of Γ is $\mathfrak{S}_3$, the symmetric group on three letters)

$^2E_6$: 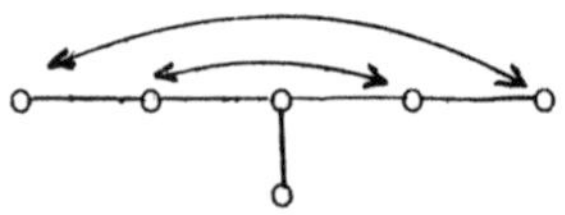

In the case of k a finite field, the Γ-diagram of type $^6D_4$ is not admissible, for it can only occur if there is a homomorphism of Γ onto $\mathfrak{S}_3$. This is impossible since Γ is cyclic and $\mathfrak{S}_3$ is not.

For more discussion on this, see Hertzig, Proc. A.M.S., vol. 12 (1961), pp. 657-660, and Dickson, "Linear Groups."

3.2 Some special cases (finite fields)

We first wish to show that if $k = \mathbb{F}_q$, a finite field, then any connected algebraic group G defined over k contains a Borel group defined over k. The proof is a direct one, and uses the results of Lang, Amer. J. Math., vol. 78 (1956), 555-563.

In order to prove the theorem below, we need to state some elementary facts on generic points. If V is an irreducible algebraic set defined over k (here k is arbitrary), then there exists a point $t = (t_1, \ldots, t_n) \in V$ such that $\dim k(t)/k = \dim V$ (see I, §1.2). Such a point $t \in V$ is called a generic point of V over k. (In general, for any $t \in V$, one has $\dim k(t)/k \leq \dim V$.) For a generic point t of V over k and for every $f \in k(V)$ (the field of k-rational functions on V) f(t) is always defined; and the mapping $f \to f(t)$ is a k-isomorphism of k(V) with k(t). (See [4].)

The main theorem due to Lang is:

Theorem 3.2.1: Let $k = \mathbb{F}_q$ be a finite field, and G a connected algebraic group defined over k. If $x \to x^{(q)}$, $x \in G$, denotes the Frobenius endomorphism of G, then the mapping $x \to x^{(q)}x^{-1}$, $x \in G$, is surjective. (See I, §2.2 for a discussion of the Frobenius endomorphism.)

Proof: Let $y \in G$; we must show there exists an $x \in G$ such that $x^{(q)}x^{-1} = y$. Put $K = k(y)$; since G is defined over k, G is also defined over K, so there exists a generic point t of G over K, and $K(t) \cong K(G)$. Let $w = t^{(q)}yt^{-1}$; we assert that w is also a generic point of G over K. For, since $y \in K$, one has $K(w,t) = K(w,t^{(q)})$, thus $K(t)/K(w)$ is a separable algebraic extension, which implies $\dim K(w)/K = \dim K(t)/K = \dim G$. Since t is a generic point of G over K, t is also a generic point of G over k; applying our result about w to the special case of $y = 1$, we see that $t^{(q)}t^{-1}$ is a generic point of G over k. Now $k(t)$ and K are independent over k (since $\dim K(t)/K = \dim k(t)/k$) so by examining the extensions below, it is easily seen that $t^{(q)}t^{-1}$ is a generic point of G over K.

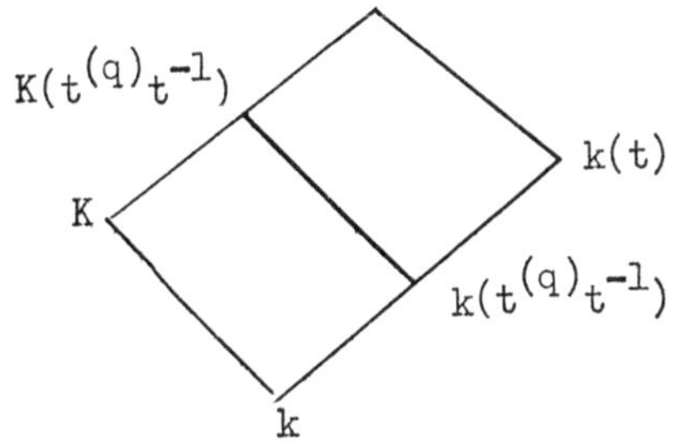

From this, we have $K(t^{(q)}t^{-1}) \cong K(G) \cong K(t^{(q)}yt^{-1})$, so there exists a K-isomorphism $\sigma : K(t^{(q)}t^{-1}) \to K(t^{(q)}yt^{-1})$ satisfying $\sigma(t^{(q)}t^{-1}) = t^{(q)}yt^{-1}$. Now $K(t^{(q)}t^{-1}) \subset K(t)$, so that extending the isomorphism σ to $K(t)$, one has $\sigma(t) = u$, where $K(u) \supset K(t^{(q)}yt^{-1})$. Since $t \to t^{(q)}t^{-1}$ is a rational function, σ satisfies $\sigma(t^{(q)}t^{-1}) = u^{(q)}u^{-1} = t^{(q)}yt^{-1}$. If we put $x = t^{-1}u$, then $x^{(q)}x^{-1} = y$.

Corollary 3.2.2: With G, k as in Theorem 3.2.1, G contains a Borel subgroup defined over k. (Thus G contains no k-compact semi-simple subgroup, and so if G is semi-simple, G is of Steinberg type.)

Proof: Let B be a Borel subgroup of G. Since the Frobenius endomorphism of G is one-to-one, surjective, and rational, the image $B^{(q)}$ of B under this endomorphism is again a Borel subgroup of G. Thus there exists an element $y \in G$ such that $B^{(q)} = yBy^{-1}$. By Theorem 3.2.1 there is an element $x \in G$ such that $y = x^{(q)}x^{-1}$; if we take $B_1 = x^{-1}Bx$, then the equation $B^{(q)} = x^{(q)}x^{-1}Bxx^{-(q)}$ shows that $B_1{}^{(q)} = B_1$. Since the Frobenius endomorphism is the generating automorphism of the Galois group $Gal(\bar{k}/k)$ (in the topological sense) it follows that B_1 is defined over k.

Corollary 3.2.3: With G, k as in Theorem 3.2.1, G contains a maximal torus T defined over k.

(The proof is similar to that of Coroll. 3.2.2.)

Corollary 3.2.4: With G, k as in Theorem 3.2.1, one has $H^1(k,G) = 1$.

Proof: If K/k is finite, then it is Galois (since k is finite), and $Gal(K/k) = \{1,(q),\ldots,(q)^{m-1}\}$, where (q) is the Frobenius endomorphism of k. If (g_σ) is a one-cocycle of Gal(K/k) in G, then by Theorem 3.2.1 $g_{(q)} = g^{(q)}g^{-1}$ for some $g \in G$. But then the cocycle condition implies $g_{(q)^i} = g^{(q)^i}g^{-1}$, so $g_\sigma = g^\sigma g^{-1}$ for all $\sigma \in \Gamma$.

Now let k be any perfect field. We wish to consider certain conditions on the field k, which are suggested by the results just obtained where k was finite. In particular, we are interested in finding fields k for which the classification problem reduces to the problem of classifying semi-simple groups of Steinberg type. We first consider the statements:

(i) There exists no k-compact semi-simple algebraic group.

(i)' All connected semi-simple algebraic groups defined over k are of

Steinberg type.

(ii) $H^1(k,G) = 1$ for every connected algebraic group G defined over k.

It is clear from what we have shown in previous sections that (i) and (i)' are equivalent. Before discussing other equivalences, we obtain two more statements.

Let k_1/k be a finite extension, and $\mathfrak{K}$ a normal division algebra over Ω which is defined over k_1 ('normal' means center $\mathfrak{K} = \Omega$, and 'division algebra defined over k_1' means that $\mathfrak{K}$ is a finite-dimensional vector space over Ω with an associative multiplication both defined over k_1 such that the set $\mathfrak{K}_{k_1}$ of k_1-rational points is a division algebra). Put $G_1 = SL(n,\mathfrak{K}) = \{g \in GL(n,\mathfrak{K}) \mid N(g) = 1\}$. (Here the norm N is the "reduced norm," i.e., $N(x) = \det M(x)$ where $x \to M(x)$ is the absolutely irreducible representation of the normal simple algebra $M_n(\mathfrak{K})$. Now G_1 is an algebraic group defined over k_1 (the linear representation given by the regular representation shows that $SL(n,\mathfrak{K})$ is a linear algebraic group). In fact, if $\dim \mathfrak{K} = r^2$, then $G_1 \cong SL(nr)$ over $\bar{k}$, the usual special linear group, so G_1 is type (A). Now put $G = R_{k_1/k}(G_1)$; then G is defined over k, and G is k-compact if and only if G_1 is k_1-compact (see end of §3.1). But G_1 is k_1-compact if and only if $n = 1$. For, if $n > 1$, it can be shown that G_1 contains a non-trivial unipotent group defined over k_1, (e.g., for $n = 2$, $\{\begin{pmatrix}1 & \xi\\ 0 & 1\end{pmatrix} \mid \xi \in \mathfrak{K}\}$), hence G_1 is not k_1-compact; on the other hand, if $n = 1$, then G_1 is just the multiplicative group of norm 1 elements in $\mathfrak{K}$, and since $\mathfrak{K}_{k_1}$ is a division algebra, $\mathfrak{K}_{k_1}$ contains no nilpotent, hence no unipotent elements, which implies G_1 is k_1-compact (Prop. 2.2.9). Thus we see that statement (i) implies the following statement:

(iii) If k_1/k is finite, then there are no non-trivial division algebras defined over k_1 (equivalently, $\mathcal{B}(k_1) = H^2(k_1, \mathbb{G}_m) = 1$).

At the end of chapter I (Example 1, §4.4) we showed for S a non-degenerate symmetric bilinear form defined over k (char. k ≠ 2) that the group G = SO(n,S) is k-compact if and only if S is k-anisotropic (this means that the polynomial $S(x) = \Sigma a_{ij}x_ix_j$ has no non-trivial zero in k). Also, it is known that if n > 2, G is semi-simple (if n = 1 or 2, G is abelian). Since S is a homogeneous polynomial of degree 2 in n variables, if one wishes (i) to be true, then for n > 2, S should have a non-trivial zero in k. This suggests the next statement on k.

(iv) (C_1) Every homogeneous polynomial in k of degree d and of m variables has a non-trivial zero in k if m > d.

We can show (iv) => (iii). Suppose $\mathfrak{K}$ is a division algebra defined over k_1, of degree r (i.e., of dimension r^2). Then N(x) = det(M(x)) is a homogeneous polynomial of r^2 variables, and degree r. If (iv) is satisfied and r > 1, then N(x) has a non-trivial zero in k, which is impossible (i.e., N(x) = 0 <=> x = 0), hence r = 1, and so (iv) => (iii).

It can also be shown that statements (i) and (ii) are equivalent; we will prove (ii) => (i), and for (i) => (ii), see [B] (Springer). The implication (iii) => (ii) was conjectured by Serre in 1962 at the Brussels Colloquium; he proved it for solvable, or classical semi-simple groups, and Steinberg proved the general case in 1965, I.H.E.S., no. 25. Thus we have (i') <=> (i) <=> (ii) <=> (iii) <= (iv). The natural conjecture that (iii) ⇒ (iv) has been disproved by Ax. (See Serre [D].) The examples of k for which the condition (iv) (hence (iii)) is satisfied are: $k = \mathbb{F}_q$, a finite field (Wedderburn); k a maximal unramified extension of a $\mathfrak{p}$-adic field; k an algebraic function field of one variable over an algebraically closed field (Tsen). See [C], Chap. X, §7 for more discussion on (iii) and (iv).

We now prove (ii) => (i). Actually, we will only need to assume that (ii) holds for all connected semi-simple algebraic groups defined over k, and of Steinberg type. Let (G,T) be any connected semi-simple algebraic group defined over k, and consider G as a k-form of the Chevalley group $(\underline{G},\underline{T})$. Let $(\varphi_\sigma) \in \mathrm{Aut}(\underline{G},\underline{T})$ be the one-cocycle corresponding to this k-form; since $\mathrm{Aut}(\underline{G},\underline{T}) = \Theta \cdot \mathrm{Inn}(\underline{G},\underline{T})$ (Prop. 1.4.1), we may write $\varphi_\sigma = \theta_\sigma \circ I_{\underline{g}_\sigma}$ where $\theta_\sigma \in \Theta$, $\underline{g}_\sigma \in N(\underline{T})$. Since $N(\underline{T})$ is normal in $\mathrm{Aut}(\underline{G},\underline{T})$ and all $\theta \in \Theta$ are k-rational (Remark after Lemma 3, §2.4), the cocycle condition on φ_σ implies that $\theta_{\sigma\tau} \circ I_{\underline{g}_{\sigma\tau}} = \theta_\sigma \circ I_{\underline{g}_\sigma{}^\tau} \circ \theta_\tau \circ I_{\underline{g}_\tau} = \theta_\sigma \circ \theta_\tau \circ I_{\underline{g}'} \circ I_{\underline{g}_\tau}$ (where $g' \in N(\underline{T})$), hence $\theta_{\sigma\tau} = \theta_\sigma\theta_\tau$ (since $\Theta \cap N(\underline{T}) = \{1\}$). Thus (θ_σ) is a one cocycle (actually a homomorphism) of Γ in Θ, and so corresponds to a k-form $(\underline{\underline{G}},\underline{\underline{T}})$ of $(\underline{G},\underline{T})$. We show that $(\underline{\underline{G}},\underline{\underline{T}})$ is of Steinberg type. Recall that Θ leaves fixed a given fundamental system Δ; let $\underline{B} = \underline{B}_\Delta$ be the Borel subgroup of $(\underline{G},\underline{T})$ corresponding to this fundamental system, and define $\underline{\underline{B}} = f_1^{-1}(\underline{B})$, where f_1 is the isomorphism of $(\underline{\underline{G}},\underline{\underline{T}})$ onto $(\underline{G},\underline{T})$ determined by (θ_σ). Since $\underline{B}$ is defined over k, $\underline{B}^\sigma = f_1^\sigma(\underline{\underline{B}}^\sigma) = \underline{B}$, for all $\sigma \in \Gamma$, thus $f_1(\underline{\underline{B}}^\sigma) = f_1 \circ f_1^{-\sigma} \circ f_1^\sigma(\underline{\underline{B}}^\sigma) = \theta_\sigma^{-1}(\underline{B}) = \underline{B}_{\theta^*_\sigma(\Delta)} = \underline{B}_\Delta = \underline{B}$, which shows $\underline{\underline{B}}^\sigma = \underline{\underline{B}}$ for all $\sigma \in \Gamma$. Thus $\underline{\underline{B}}$ is defined over k, and so $\underline{\underline{G}}$ is of Steinberg type. Now if f is the isomorphism of (G,T) onto $(\underline{G},\underline{T})$ determined by (φ_σ), then the mapping $f_2 = f_1^{-1} \circ f$ is an isomorphism of (G,T) onto $(\underline{\underline{G}},\underline{\underline{T}})$ which satisfies $f_2^\sigma \circ f_2^{-1} = (f_1^{-1} \circ f)^\sigma \circ f^{-1} \circ f_1 = f_1^{-\sigma} \circ \theta_\sigma \circ I_{\underline{g}_\sigma} \circ f_1 = f_1^{-1} \circ I_{\underline{g}_\sigma} \circ f_1 = I_{f_1^{-1}(\underline{g}_\sigma)}$. Put $\underline{\underline{g}}_\sigma = f_1^{-1}(\underline{g}_\sigma)$; clearly $I_{\underline{\underline{g}}_\sigma}$ is an element in $H^1(k, \underline{\underline{G}}/\underline{\underline{Z}})$. If $I_{\underline{\underline{g}}_\sigma} \sim 1$, then G is k-isomorphic to $\underline{\underline{G}}$, hence G is of Steinberg type.

Thus we see that (ii) => (i).

Remark: Our argument above shows that every connected semi-simple algebraic group G defined over k is obtained in two 'stages' from a (uniquely determined) Chevalley group $(\underline{G},\underline{T})$. First the Chevalley group $\underline{G}$ is 'twisted' by outer automorphisms in Θ to obtain a Steinberg group $(\underline{\underline{G}},\underline{\underline{T}})$, then $\underline{\underline{G}}$ is 'twisted' by inner automorphisms to obtain G. In other words, one has

$$\begin{array}{ccc} (G,T) & \xrightarrow{\ f\ } & (\underline{G},\underline{T}) \\ & {}_{f_2}\searrow \qquad \nearrow{}_{f_1} & \\ & (\underline{\underline{G}},\underline{\underline{T}}) & \end{array}$$

is a commutative diagram of K-isomorphisms (K/k is finite), where $f_1 \leftrightarrow (\theta_\sigma)$, $\theta_\sigma \in \Theta$ (a group of outer automorphisms of $(\underline{G},\underline{T})$), and $f_2 \leftrightarrow (I_{\underline{\underline{g}}_\sigma})$, $\underline{\underline{g}}_\sigma \in N(\underline{\underline{T}})$. We also note that in the identification of X with $\underline{X}$ (via f*), that the following actions of Γ on X and $\underline{X}$ become identified: $\sigma^{-1} = \varphi_\sigma^*$, $[\sigma]^{-1} = \theta_\sigma^*$, $w_\sigma^{-1} = I_{\underline{\underline{g}}_\sigma}^*$.

3.3 The invariant $\gamma(G)$ and the classification over a $\mathfrak{p}$-adic field[13) ([11], Part II)

Let G be a connected semi-simple algebraic group defined over k, and let all notations be as in the proof of (ii) => (i) at the end of the last section. Thus, if (G,f) is a k-form of the Chevalley group $(\underline{G},\underline{T})$, then $f \leftrightarrow (\varphi_\sigma)$, $\varphi_\sigma \in \mathrm{Aut}(\underline{G},\underline{T})$, $\varphi_\sigma = \theta_\sigma \circ I_{\underline{\underline{g}}_\sigma}$, and (θ_σ) is a one-cocycle of Γ in Θ which determines a k-form $(\underline{G},f_1)$ of $\underline{G}$, and $\underline{\underline{G}}$ is a Steinberg group. The isomorphism $f_2 = f_1^{-1} \circ f$ of (G,T) onto $(\underline{\underline{G}},\underline{\underline{T}})$ corresponds to an inner automorphism $I_{\underline{\underline{g}}_\sigma}$ of $(\underline{\underline{G}},\underline{\underline{T}})$. The isomorphism f_2^{-1} of $(\underline{\underline{G}},\underline{\underline{T}})$ onto (G,T) corresponds also to an inner automorphism I_{g_σ} of (G,T), where $g_\sigma = f_2^{-1}(\underline{\underline{g}}_\sigma^{-1})$. Since I_{g_σ} determines g_σ modulo Z, the center of

G, the cocycle condition on I_{g_σ} implies $g_\sigma {}^\tau g_\tau g_{\sigma\tau}^{-1} = c_{\sigma,\tau} \in Z$, hence $(c_{\sigma,\tau}) \in H^2(\Gamma, Z)$. We wish to show that the class of $(c_{\sigma,\tau})$ is uniquely determined by G, that is, it is independent of the choice of the isomorphisms f_1, f_2.

We have already noted that G determines (up to k-isomorphism) the Chevalley group $\underline{G}$ of which G is a k-form. We now show that the Steinberg group $\underline{\underline{G}}$ obtained from $\underline{G}$ by the method just described is also determined up to k-isomorphism. Thus, let f' be another isomorphism of (G,T) onto $(\underline{G},\underline{T})$, and let $\underline{\underline{G}}'$, f_1', f_2' $(f' = f_1' \circ f_2')$ be the Steinberg group and isomorphisms obtained in the same manner as before (that is, $G \xrightarrow{f_2'} \underline{\underline{G}}'$, $\underline{\underline{G}}' \xrightarrow{f_1'} \underline{G}$, with $(\varphi_\sigma') \longleftrightarrow f'$, $\varphi_\sigma' = \theta_\sigma' \circ I_{\underline{g}_\sigma'}$, $(\theta_\sigma') \longleftrightarrow f_1'$, $(I_{g_\sigma'}) \longleftrightarrow f_2'$ where $\underline{\underline{g}}_\sigma' = f_1'^{-1}(\underline{g}_\sigma'))$. Now the isomorphisms f' and f differ only by an automorphism of $(\underline{G},\underline{T})$, hence $f' = \theta \circ I_{\underline{g}} \circ f$, where $\theta \in \Theta$, $\underline{g} \in N(\underline{T})$. Since θ is defined over k and $N(\underline{T})$ is normal in $\mathrm{Aut}(\underline{G},\underline{T})$, one has $\theta_\sigma' \circ I_{\underline{g}_\sigma'} = \varphi_\sigma' = f'^\sigma \circ f'^{-1} = \theta \circ I_{\underline{g}^\sigma} \circ f^\sigma \circ f^{-1} \circ I_{\underline{g}}^{-1} \circ \theta^{-1}$ $= \theta \circ I_{\underline{g}^\sigma} \circ \theta_\sigma \circ I_{\underline{g}_\sigma} \circ I_{\underline{g}}^{-1} \circ \theta^{-1} = \theta \circ \theta_\sigma \circ \theta^{-1} \circ I_{\underline{g}''}$ for some $\underline{g}'' \in N(\underline{T})$, hence $\theta_\sigma' = \theta \circ \theta_\sigma \circ \theta^{-1}$ (since $\Theta \cap N(\underline{T}) = \{1\}$). Thus the 1-cocycles (θ_σ') and (θ_σ) are cohomologous, which implies $\underline{\underline{G}}'$ is k-isomorphic to $\underline{\underline{G}}$ (I, Coroll. 3.1.4'). More precisely, there exists a k-isomorphism h, which makes the following diagram commute:

$$(\star) \qquad \begin{array}{ccccc} G & \xrightarrow{f_2'} & \underline{\underline{G}}' & \xrightarrow{f_1'} & \underline{G} \\ {\scriptstyle I_g}\uparrow & & \uparrow{\scriptstyle h} & & \uparrow{\scriptstyle \theta} \\ G & \xrightarrow{f_2} & \underline{\underline{G}} & \xrightarrow{f_1} & \underline{G} \end{array}$$

Now $f'^{-1} \circ \theta \circ f = f^{-1} \circ I_{\underline{g}}^{-1} \circ \theta^{-1} \circ \theta \circ f = I_{f^{-1}(\underline{g}^{-1})}$; so if we set $g = f^{-1}(\underline{g}^{-1})$, then the inner automorphism I_g of G makes the diagram commute.

Let $(I_{g_\sigma'}) \longleftrightarrow f_2'^{-1}$; then since $f_2' = h \circ f_2 \circ I_g^{-1}$ and h is defined over k, we have $I_{g_\sigma'} = f_2'^{-\sigma} \circ f_2' = I_{g^\sigma} \circ f_2^{-\sigma} \circ h^{-\sigma} \circ h \circ f_2 \circ I_g^{-1} = I_{g^\sigma} \circ I_{g_\sigma} \circ I_g^{-1}$. Thus $g_\sigma' = c_\sigma g^\sigma g_\sigma g^{-1}$, where $c_\sigma \in Z$; if $c'_{\sigma,\tau}$ denotes the coboundary of g_σ', then this equation implies that $c'_{\sigma,\tau} = \delta(c_\sigma)c_{\sigma,\tau}$. Thus the cohomology class of $(c_{\sigma,\tau})$ is uniquely determined by the group G.

Definition: The cohomology class of the 2-cocycle $(c_{\sigma,\tau})$ determined by G will be denoted $\gamma(G)$. (Thus $\gamma(G)$ is an element in $H^2(\Gamma,Z)$ uniquely determined by G.)

The element $\gamma(G)$ is a generalization of the Hasse invariant, as shown in the following example.

Example: (Cf. Ex. 4 of I, §3.2) Let $\mathfrak{K}$ be a normal division algebra of degree m over k, and let $G = SL(n,\mathfrak{K})$. Then G is $\bar{k}$-isomorphic to SL(nm), so we can take $\underline{G} = SL(nm)$. There exists an isomorphism $f: \mathfrak{K} \to \mathfrak{M}_m$, the total matrix algebra over $\bar{k}$ of degree m, and f can be extended to an isomorphism $f: \mathfrak{M}_n(\mathfrak{K}) \to \mathfrak{M}_{nm}$. Since $G \subset \mathfrak{M}_n(\mathfrak{K})$ and $\underline{G} \subset \mathfrak{M}_{nm}$, f is an isomorphism of G onto $\underline{G}$, and f is defined over K where K/k is finite. If $\varphi_\sigma = f^\sigma \circ f^{-1}$ is considered as an automorphism of $\mathfrak{M}_{nm}$, then φ_σ is inner (Skolem-Noether); put $\varphi_\sigma = I_{\phi_\sigma}$ (note one can take $\phi_\sigma \in \underline{G}$), then the cocycle condition on φ_σ implies $\phi_\sigma^\tau \phi_\tau = \lambda_{\sigma,\tau}\phi_{\sigma\tau}$ where $(\lambda_{\sigma,\tau})$ is a 2-cocycle of Γ in $\bar{k}$. The inverse of the class of the cocycle $(\lambda_{\sigma,\tau})$ in $H^2(k, \mathbb{G}_m)$ is denoted $c(\mathfrak{M}_n(\mathfrak{K}))$ and called the "Hasse invariant" of $\mathfrak{M}_n(\mathfrak{K})$. Since $\varphi_\sigma = I_{\phi_\sigma}$, we see that the Steinberg group $\underline{\underline{G}}$ corresponding to (G,f) coincides with $\underline{G}$, thus using the previous notation, $f = f_2$, $g_\sigma = f^{-1}(\phi_\sigma^{-1})$, and $c_{\sigma,\tau} = g_\sigma^\tau g_\tau g_{\sigma\tau}^{-1} = f^{-1}(\lambda_{\sigma,\tau}^{-1})$. Thus $f(c_{\sigma,\tau}) = \lambda_{\sigma,\tau}^{-1}$, so that in this case, through the isomorphism induced by f, $\gamma(G)$ coincides with the Hasse invariant $c(\mathfrak{M}_n(\mathfrak{K}))$.

If one applies a similar argument to the spin group (i.e., the simply connected covering group of the special orthogonal group), it can be seen that $\gamma(G)$ is a generalization of the Minkowski invariant.

The theorem 3.3.2 below states that when k is a $\wp$-adic field, $\gamma(G)$ is a complete invariant of G, that is, the k-isomorphism class of G (for G simply connected) is determined by $\gamma(G)$. We will use this invariant, then, to solve the classification problem for k a $\wp$-adic field.

A key fact used in the proof of Theorem 3.3.2 below is the following:

<u>Theorem 3.3.1</u>: Let k be a $\wp$-adic field, and G a connected simply connected semi-simple algebraic group defined over k. Then $H^1(k,G) = 1$. (Cf. M. Kneser, "Galois-Kohomologie halbeinfacher algebraischer Gruppen über $\wp$-adischen Körpern," II, Math. Z., 89 (1965), 250-272. Recently, Bruhat and Tits gave a new proof of this theorem, using an existence theorem of "Iwahori subgroups." See Compte Rendu, Paris 263 (1966).)

<u>Theorem 3.3.2</u>: Let k be a $\wp$-adic field, and G, G' simply connected semi-simple algebraic groups defined over k. If there is a K-isomorphism $\varphi: G \to G'$ (where K/k is finite Galois) such that $\varphi^{-\sigma} \circ \varphi \in \mathrm{Inn}(G)$ and $\varphi^{\#}(\gamma(G)) = \gamma(G')$, then G and G' are k-isomorphic. ($\varphi^{\#}$ is the isomorphism of $H^2(\Gamma,Z) \to H^2(\Gamma,Z')$ induced by the Γ-isomorphism $\varphi|Z\colon Z \to Z'$.)

<u>Proof</u>: By our remarks earlier in this section, we may assume that G and G' are obtained by twisting from the same Steinberg group $\underline{\underline{G}}$, i.e., we may assume the following diagram commutes:

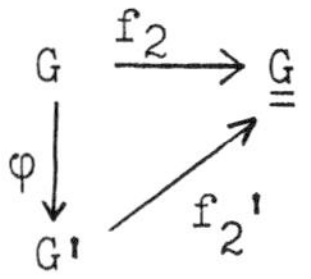

Now $f_2^{-\sigma} \circ f_2 = I_{g_\sigma}$ and $f_2'^{-\sigma} \circ f_2' = I_{g_\sigma'}$ where $g_\sigma \varepsilon G$, $g_\sigma' \varepsilon G'$, so that $\varphi^{-\sigma} \circ \varphi = f_2^{-\sigma} \circ f_2'^{\sigma} \circ \varphi = f_2^{-\sigma} \circ f_2 \circ \varphi^{-1} \circ f_2'^{-1} \circ f_2'^{\sigma} \circ \varphi = I_{g_\sigma \varphi^{-1}(g_\sigma')^{-1}}$; this implies $(g_\sigma \varphi^{-1}(g_\sigma')^{-1}) \varepsilon H^1(\Gamma, G/Z)$. Since $\varphi^{\#}(\gamma(G)) = \gamma(G')$, we may assume (modifying g_σ' if necessary) that $\varphi(c_{\sigma,\tau}) = c'_{\sigma,\tau}$, where $c_{\sigma,\tau} = \delta(g_\sigma)$, $c'_{\sigma,\tau} = \delta(g_\sigma')$; then using the relation $\varphi^{-\tau} \circ \varphi = I_{g_\tau \varphi^{-1}(g_\tau')^{-1}}$, one has $\delta(g_\sigma \varphi^{-1}(g_\sigma')^{-1}) = c_{\sigma,\tau}\varphi^{-1}(c'_{\sigma,\tau})^{-1} = 1$. Since $H^1(\Gamma,G) = 1$ (Theorem 3.3.1), the exactness of the sequence $H^1(\Gamma,G) \to H^1(\Gamma,G/Z) \xrightarrow{\delta} H^2(\Gamma,Z)$ then implies that the one-cocycle $(g_\sigma \varphi^{-1}(g_\sigma')^{-1})$ is cohomologous to (1), thus there is an element $g \varepsilon G$ such that $g_\sigma \varphi^{-1}(g_\sigma')^{-1} = g^\sigma g^{-1}$ for all $\sigma \varepsilon \Gamma$. But then $\varphi^{-\sigma} \circ \varphi = I_g^{\ \sigma} I_g^{\ -1}$, and so $\varphi \circ I_g$ is a k-isomorphism of G onto G'.

<u>Remark</u>: The classification of semi-simple algebraic groups over a $\mathcal{L}$-adic field k is reduced to the classification of Steinberg groups defined over k and the classification of elements of the form $f_2^{\#}(\gamma(G)) \varepsilon H^2(\Gamma,Z)$ as follows. We have seen that any connected semi-simple algebraic group G defined over k is obtained from a uniquely determined (up to k-isomorphism) Steinberg group $\underline{\underline{G}}$ by twisting by an inner automorphism (i.e., f_2: $G \to \underline{\underline{G}}$ where $f_2^{-\sigma} \circ f_2$ is inner). Now f_2 is not unique, but if f_2':$G \to \underline{\underline{G}}$ is another isomorphism such that $f_2'^{-\sigma} \circ f_2'$ is inner, then $f_2' = \underline{\underline{\theta}} \circ f_2 \circ I_g$ where $g \varepsilon G$ and $\underline{\underline{\theta}}$ is a k-automorphism of $\underline{\underline{G}}$ (refer to the diagram (☘) earlier in this section, and replace f_1' by f_1, and replace h by $\underline{\underline{\theta}}$). Since $f_2'^{\#} = \underline{\underline{\theta}}^{\#} \circ f_2^{\#}$, we see that $f_2^{\#}(\gamma(G))$ is uniquely determined modulo the action of $\underline{\underline{\theta}} \varepsilon \underline{\underline{\Theta}}_k$ (where $\underline{\underline{\Theta}} = f_1^{-1} \Theta f_1$). Thus Theorem 3.3.2 shows that the k-isomorphism class of G is uniquely determined by the Steinberg group $\underline{\underline{G}}$ and by the element $f_2^{\#}(\gamma(G))$, modulo the operation of $\underline{\underline{\Theta}}_k$.

Throughout the following discussion, k and G are as in Theorem 3.3.2, and Z is the center of G. We are interested in determining $H^2(k,Z)$. If $\underline{\underline{G}}$ is the Steinberg group from which G is obtained, the center $\underline{\underline{Z}}$ of $\underline{\underline{G}}$ may be identified with Z. If $\hat{Z}$ denotes the character group of Z, then $\hat{Z} = \mathrm{Hom}(Z, \mathbb{G}_m)$ and $X = \mathrm{Hom}(T, \mathbb{G}_m)$, so there is a canonical homomorphism of X into $\hat{Z}$ whose kernel is the set of characters of T which annihilate Z. Thus via this homomorphism, one may identify $\hat{Z} = X/\{\vee\}_{\mathbb{Z}}$. By Tate duality, it is known that $H^2(k,Z) \cong H^0(k,\hat{Z}) = \hat{Z}^{\Gamma}$ (see, e.g., [C]). Thus we only need to determine the Γ-invariant elements of $\hat{Z} = X/\{\vee\}_{\mathbb{Z}}$. It should be noted that the [ ]-action of Γ is just the usual action of Γ on $X/\{\vee\}_{\mathbb{Z}}$, for these two actions differ only by an element in the Weyl group, and the Weyl group acts trivially on $X/\{\vee\}_{\mathbb{Z}}$. Thus we wish to determine $\hat{Z}^{[\Gamma]}$. The table below lists the simply connected groups having non-trivial center and the corresponding groups $\mathrm{Aut}(X, \Delta)$ (in the table, (n) denotes the cyclic group of order n). Of course, if the center of G is trivial, then Theorem 3.3.2 implies G is of Chevalley type, so there is no classification problem.

(††)

G	Z ($\cong \hat{Z}$)	$\Theta \cong \mathrm{Aut}(X,\Delta)$
A_ℓ	$(\ell + 1)$	(2)
B_ℓ, C_ℓ	(2)	(1)
D_ℓ ℓ even	$(2) \times (2)$	(2) $(\ell \neq 4)$; $\mathfrak{S}_3$ $(\ell = 4)$
D_ℓ ℓ odd	(4)	(2)
E_6	(3)	(2)
E_7	(2)	(1)

Since Θ is a group of outer automorphisms, it can be considered as an automorphism group of Z. Its action on Z is as follows: when $\Theta = \{1\}$,

Θ acts trivially on Z; when $\Theta = \{1, \theta_0\}$, then $\theta_0 : z \to z^{-1}$ for all $z \in Z$; when $Z = (2) \times (2)$ and $\Theta = \mathfrak{S}_3$, Θ acts on Z in such a way that Θ permutes the three non-identity elements of Z arbitrarily. Finally, we note that if $\hat{Z}^{[\Gamma]} = (1)$ or (2), then $\underline{\underline{\Theta}}_k$ operates trivially on $H^2(k, \underline{\underline{Z}})$, and in all other cases, Γ operates trivially on Z, so that $\underline{\underline{G}} = \underline{G}$, $\underline{\underline{Z}} = \underline{Z}$, $\underline{\underline{\Theta}}_k = \Theta$. In the latter case, $H^2(k, \underline{\underline{Z}}) = H^2(k, \underline{Z})$ is dual to $H^0(k, \hat{\underline{Z}}) = \hat{\underline{Z}}$, and the operation of Θ on $\hat{\underline{Z}}$ is determined by the above.

When k is $\mathfrak{p}$-adic, it is also known that $H^2(k, \mathbb{G}_m) = \mathcal{B}(k) \cong \mathbb{Q}/\mathbb{Z}$ (this is a special case of Tate duality). If one fixes this isomorphism, then for each integer $m > 0$, there exists a normal division algebra $\mathfrak{K}_m$ of degree m over k such that $\mathfrak{K}_m \longleftrightarrow \frac{1}{m}$; and all normal division algebras of degree m over k are given by $\mathfrak{K}_m^{\nu} \longleftrightarrow \frac{\nu}{m}$ with $(\nu, m) = 1$. In this case, $SL(1, \mathfrak{K}_m^{\nu}) =$ (set of norm 1 elements of $\mathfrak{K}_m^{\nu}$), and this group is k-compact, having Γ-diagram

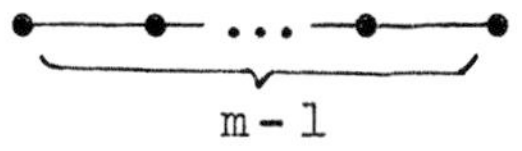

It will be seen later that in fact, this is the only k-compact group for k $\mathfrak{p}$-adic. We now merely list all k-rank = 1 classical groups for k a $\mathfrak{p}$-adic field; it will turn out that all Γ-diagrams of semi-simple connected algebraic groups defined over k can be composed from these (see §3.1). In the chart, $\mathbb{F}$ denotes a quaternion form, S a symmetric form, a zero subscript denotes an anisotropic form (i.e., $\mathbb{F}_0, S_0$), and superscript (n) denotes a form in n variables (i.e., $\mathbb{F}^{(n)}$, $S^{(n)}$).

Below the chart, we list (beginning with the group $SU(4, \mathbb{F}^{(4)}, k'/k)$ and following down the chart in order) the k-compact groups which correspond to the connected sub-diagrams of black dots (when there are two such sub-diagrams of a Γ-diagram, we refer to the one on the right).

k-rank = 1	absolutely simple classical groups, k a $\mathfrak{p}$-adic field
$SL(2, \mathfrak{K}_m^{\nu})$	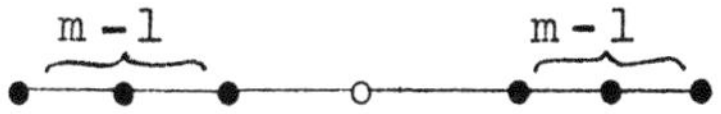
$SU(3,F^{(3)},k'/k)$, $F^{(3)}$:hermitian, $F^{(3)} \sim F_o^{(1)}$, $[k':k] = 2$	
$SU(4,F^{(4)},k'/k)$, $F^{(4)}$:hermitian, $F^{(4)} \sim F_o^{(2)}$, $[k':k] = 2$	
$SO(5,S^{(5)})$, $S^{(5)} \sim S_o^{(3)}$	
$SU(3, \mathbb{F}^{(3)}, \mathfrak{K}_2)$, $\mathbb{F}^{(3)}$:hermitian, $\mathbb{F}^{(3)} \sim \mathbb{F}_o^{(1)}$	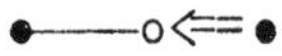
$SU(4, \mathbb{F}^{(4)}, \mathfrak{K}_2)$, $\mathbb{F}^{(4)}$:skew-hermitian, $\mathbb{F}^{(4)} \sim \mathbb{F}_o^{(2)}$	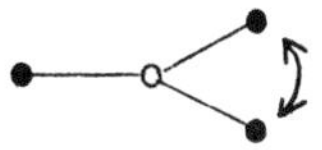
$SU(5, \mathbb{F}^{(5)}, \mathfrak{K}_2)$, $\mathbb{F}^{(5)}$:skew-hermitian, $\mathbb{F}^{(5)} \sim \mathbb{F}_o^{(3)}$	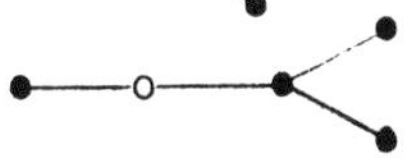

Corresponding k-compact groups

$SU(2,F_o^{(2)}) \cong SL(1, \mathfrak{K}_2)$

$SO(3,S_o^{(3)}) \sim SL(1, \mathfrak{K}_2)$ ($\sim$ means isogeneous)

$SU(1, \mathbb{F}_o^{(1)}) \cong SL(1, \mathfrak{K}_2)$

$SU(2, \mathbb{F}_o^{(2)}) \sim R_{k'/k}(SL(1, \mathfrak{K}_2'))$, where $\mathfrak{K}_2'$ is a quaternion algebra over a quadratic extension k' of k

$SU(3, \mathbb{F}_o^{(3)}) \sim SL(1, \mathfrak{K}_4)$ ($\cong SL(1, \mathfrak{K}_4^3)$).

These isogenies are obtained by considering the corresponding spin groups.

Before giving the complete classification table of connected (simply connected) semi-simple algebraic groups over a $\mathfrak{p}$-adic field, we need to explain some notation. From what we have previously shown, we only need to classify all connected, absolutely simple groups of Steinberg type over k, and then determine all possible connected, absolutely simple groups G defined over k arising from a fixed Steinberg group $\underline{\underline{G}}$ in the manner we have described. Thus, suppose we have as before,

$$\begin{array}{ccccc} G & \xrightarrow{f_2} & \underline{\underline{G}} & \xrightarrow{f_1} & \underline{G} \\ \updownarrow & & \updownarrow & & \updownarrow \\ (X,\Delta,\Delta_0,[\sigma]) & & (X,\Delta,\emptyset,[\sigma]) & & (X,\Delta). \end{array}$$

Define $\Gamma_0 = \{\sigma \in \Gamma | [\sigma] = 1\}$, and let K_0 be the fixed field of Γ_0. Then Γ_0 and K_0 are independent of the choice of the Γ-fundamental system Δ, for if Δ' is another Γ-fundamental system, there exists an element $w \in W^{\Gamma}$ such that $w\Delta = \Delta'$, which implies $[\sigma]' = w[\sigma]w^{-1}$ (where, by definition, $\chi^{[\sigma]'} = {w'_\sigma}^{-1} \chi^{\sigma}$, with $w'_\sigma(\Delta') = \Delta'^{\sigma}$); thus $[\sigma] = 1$ if and only if $[\sigma]' = 1$. (The field K_0 is called the "nuclear field" of G by T. Ono, Nagoya Jour., vol. 27, 1966.) We will denote $d = [K_0 : k]$, and use Tits' notation to denote the Dynkin diagram of $\underline{\underline{G}}$ by ${}^dX_\ell$, where $\ell = \#\Delta$ (if always d = 1, we omit it). We also say that G is of type ${}^dX_\ell$. The classification of Steinberg groups over k a $\mathfrak{p}$-adic field is as follows:

$${}^1A_\ell,\ {}^2A_\ell,\ B_\ell,\ C_\ell,\ {}^1D_\ell,\ {}^2D_\ell,\ {}^3D_4,\ {}^6D_4,\ {}^1E_6,\ {}^2E_6,\ E_7,\ E_8,\ F_4,\ G_2.$$

Now from the table of k-rank = 1 simply connected classical groups, one can compose (by Prop. 3.1.2) all possible admissible diagrams arising from these. We give this table on the next page, and then show that this, in fact, is the complete classification table of connected (simply connected) semi-simple algebraic groups defined over k.

Connected, absolutely simple algebraic groups over k, a $\mathfrak{p}$-adic field

${}^1A_\ell$: $(SL(r+1, \bar{k}_m))$

(compose the first Γ-diagram r times with itself: $\ell + 1 = m(r+1)$)

${}^2A_\ell$: ℓ even

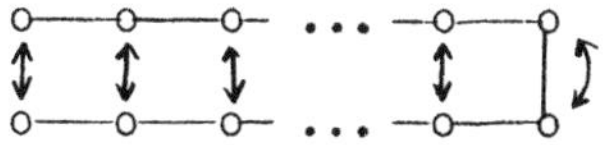

${}^2A_\ell$: ℓ odd

$(SU(\ell+1, F, k'/k))$

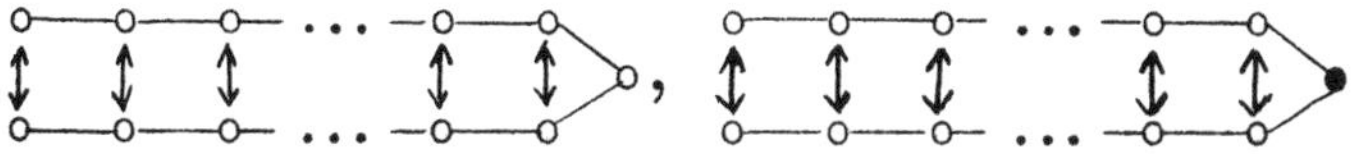

$(F \sim 0)$ $\qquad$ $(F \sim F_0^{(1)})$

B_ℓ: $(SO(2\ell+1, S))$

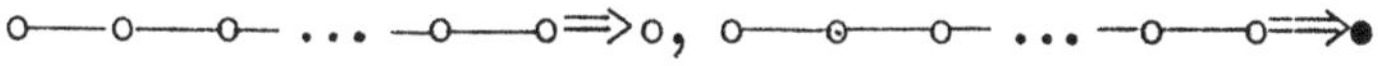

$(S \sim S_0^{(1)})$ $\qquad$ $(S \sim S_0^{(3)})$

C_ℓ: $(Sp(\ell))$

C_ℓ: ℓ even

$(SU(\ell, \mathbb{F}, \bar{k}_2))$

(every other dot black)

$(\mathbb{F} \sim 0)$

C_ℓ: ℓ odd

$(SU(\ell, \mathbb{F}, \bar{k}_2))$

(every other dot black)

$(\mathbb{F} \sim \mathbb{F}_0^{(1)})$

${}^1D_\ell$: $(SO(2\ell, S))$

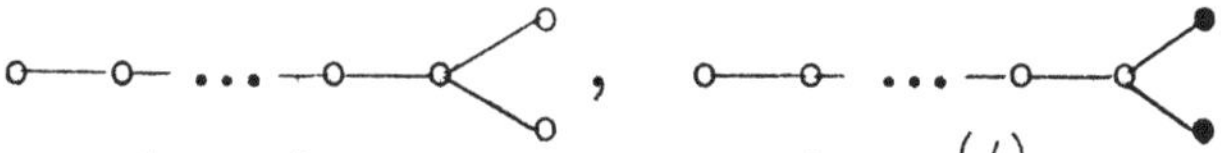

$(S \sim 0)$ $\qquad$ $(S \sim S_0^{(4)})$

${}^1D_\ell$: ℓ even

$(SU(\ell, \mathbb{F}, \bar{k}_2))$

$(\mathbb{F} \sim 0)$

${}^1D_\ell$: ℓ odd

$(SU(\ell, \mathbb{F}, \bar{k}_2))$

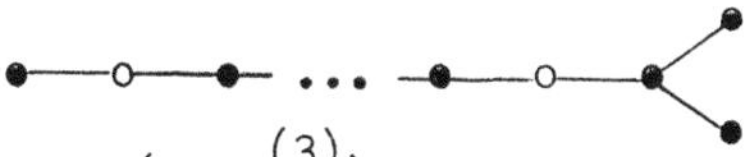

$(\mathbb{F} \sim \mathbb{F}_0^{(3)})$

${}^2D_\ell$: $(SO(2\ell, S))$

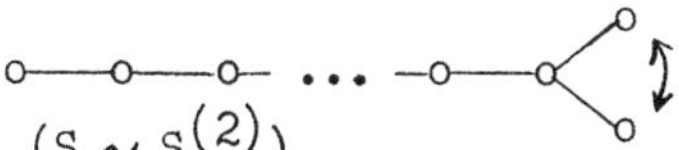

$(S \sim S_0^{(2)})$

${}^2D_\ell$: ℓ even

$(SU(\ell, \mathbb{F}, \bar{k}_2))$

$(\mathbb{F} \sim \mathbb{F}_0^{(2)})$

${}^2D_\ell$: ℓ odd

$(SU(\ell, \mathbb{F}, \bar{k}_2))$

$(\mathbb{F} \sim \mathbb{F}_0^{(1)})$

$^3D_4$: 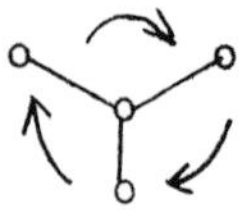

$^6D_4$: 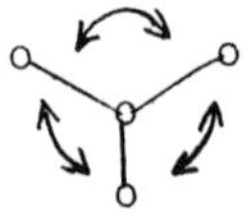 (full action of $\mathfrak{S}_3$ on 'outside' dots)

$^1E_6$: 

$^2E_6$: 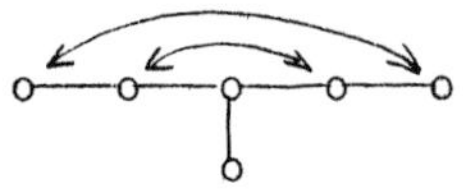

E_7: 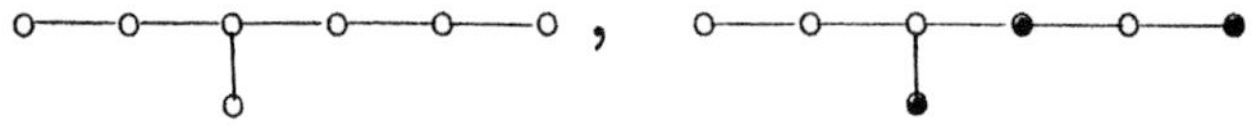

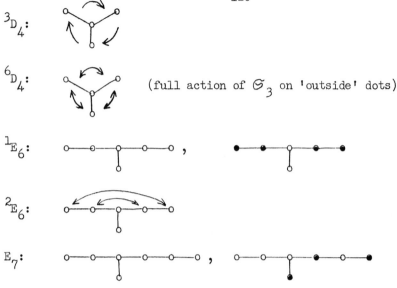

E_8, F_4, G_2: all Chevalley type.

Now we have shown previously that a connected, simply connected semi-simple algebraic group G is uniquely determined by the Steinberg group $\underline{\underline{G}}$ from which it arises, and the element $f_2^{\#}(\gamma(G))$ modulo the action of $\underline{\underline{\Theta}}_k$. Thus we make a table below of Steinberg groups $\underline{\underline{G}}$ and the possible k-isomorphism classes of G arising from each $\underline{\underline{G}}$; we see that the number of isomorphism classes is exactly the same as the number displayed on the previous chart, hence the previous chart gives the complete classification! (The chart, p.121, is an expanded version of (**), using the same notations; Z and $\underline{\underline{Z}}$ are identified.)

Some immediate consequences of the classification results over a $\mathfrak{l}$-adic field k are given below.

1. There exist no k-compact simple groups other than $SL(1, \mathfrak{K}_m^{\nu})$.

   1.1 The known k-compact groups are isomorphic to $SL(1,\mathfrak{K})$, for some normal division algebra $\mathfrak{K}$. In particular, these include groups of anisotropic quadratic (resp. hermitian, skew-hermitian) forms.

$\underline{\underline{G}}$	$\hat{Z}$	$\hat{Z}^{\Gamma}$	the action of $\underline{\underline{\Theta}}_k$	number of k-isomorphism classes arising from $\underline{\underline{G}}$
${}^1A_\ell$	$(\ell+1)$, $m \mid \ell+1$	$(\ell+1)$	$\Theta = (2)$, $z \to z^{-1}$	$\left[\frac{\ell+3}{2}\right]$
${}^2A_\ell$	$(\ell+1)$	(1) (ℓ even) (2) (ℓ odd)	trivial	1 (ℓ even) 2 (ℓ odd)
B_ℓ, C_ℓ	(2)	(2)	trivial	2
${}^1D_\ell$, (ℓ odd)	(4)	(4)	$\Theta = (2)$, $z \to z^{-1}$	3
${}^1D_\ell$, (ℓ even, > 4)	$(2) \times (2)$	$(2) \times (2)$	$\Theta = (2)$, $\{1, a \leftrightarrow b, ab\}$	3
$\ell = 4$	$(2) \times (2)$	$(2) \times (2)$	$\Theta = \mathfrak{S}_3$, $\{1, a, b, ab\}$ ($a \leftrightarrow b$, $b \leftrightarrow ab$, $a \leftrightarrow ab$)	2
${}^2D_\ell$, (ℓ even, ≥ 4)	$(2) \times (2)$	(2)	trivial	2
(ℓ odd)	(4)	(2)	trivial	2
${}^3D_4$, ${}^6D_4$	$(2) \times (2)$	(1)	trivial	1
E_6	(3)	(3)	$\Theta = (2)$, $z \to z^{-1}$	2
${}^2E_6$	(3)	(1)	trivial	1
E_7	(2)	(2)	trivial	2
E_8, F_4, G_2	(1)	(1)	trivial	1

1.2 There exist no anisotropic quadratic (resp., quaternion skew hermitian, hermitian, quaternion hermitian) forms of greater than 4 (resp., 3, 2, 1) variables.

1.3 There exist no k-compact exceptional groups.

2. The k-forms of classical groups (except D_4) are in one-to-one correspondence with k-forms of semi-simple algebras defined over k with

involution, hence are in one-to-one correspondence with multiplicative equivalence classes of ε-hermitian forms.

Weil has proved this for char. k = 0 in Jour. Ind. Math. Soc., vol. 24 (1960). More explicitly, if $(\mathfrak{A}, \iota)$ is a semi-simple algebra defined over k with involution ι, then $(\mathfrak{A}, \iota)$ can be decomposed into the direct sum of simple components and the classification of such $\mathfrak{A}$ can be reduced to the case where $\mathfrak{A}$ is simple, i.e., $\mathfrak{A}$ has one or two absolutely simple components. In the classification, three cases are considered as follows. The first case is where $\iota|\mathrm{center}(\mathfrak{A}) = \mathrm{identity}$ (ι is of the 'first kind'). Then $\mathfrak{A} = \mathfrak{M}_n(\mathfrak{k})$ is a normal simple algebra defined over k, and in the Brauer group, $\mathfrak{A}^2 \sim \mathfrak{k}^2 \sim 1$. When k is $\mathfrak{p}$-adic, this implies $\mathfrak{k} = k$ or $\mathfrak{k} = \mathfrak{k}_2$, and the involution ι is given by $a \to a^{\iota} = \mathbb{F}^{-1}\,{}^{t}\bar{a}\,\mathbb{F}$, where $\mathbb{F}$ is a symmetric or alternating matrix if $\mathfrak{k} = k$, and quaternion skew-hermitian or hermitian if $\mathfrak{k} = \mathfrak{k}_2$, and for $a = (a_{ij})$ we put ${}^{t}\bar{a} = (\bar{a}_{ji})$, the bar denoting the canonical involution if $\mathfrak{k} = \mathfrak{k}_2$, and the identity if $\mathfrak{k} = k$; $\mathbb{F}$ corresponds to a form in n variables over $\mathfrak{k}$ defined by $\mathbb{F}(x,y) = {}^{t}\bar{x}\,\mathbb{F}y$. Thus one can define the group $G = SU(\mathbb{F}) = \{g \in \mathfrak{A} \mid g^{\iota}g = 1, N(g) = 1\}$. The second case is where $\iota|\mathrm{center}(\mathfrak{A}) \neq \mathrm{identity}$ and $\mathrm{center}(\mathfrak{A}) = k \oplus k$, $\iota(x \oplus y) = y \oplus x$. In this case, $\mathfrak{A} = \mathfrak{A}_1 \oplus \mathfrak{A}_2, \mathfrak{A}_1 = \mathfrak{M}_n(\mathfrak{k})$, $\mathfrak{A}_2 = \mathfrak{M}_n(\mathfrak{k}^{-1})$, and the group G defined above is $SL(n,\mathfrak{k})$. The third case is where $\iota|\mathrm{center}(\mathfrak{A}) \neq \mathrm{identity}$ and $\mathrm{center}(\mathfrak{A}) =$ quadratic extension k' of k. In this case, $\mathfrak{A} = \mathfrak{M}_n(\mathfrak{k}')$ and the group G defined above is $SU(F,\mathfrak{k}')$ where $\mathbb{F} = \mathbb{F}^{(n)}$ is a $\mathfrak{k}'$-hermitian form with respect to an involution of the second kind on $\mathfrak{k}'$. When k is a $\mathfrak{p}$-adic field, one has always $\mathfrak{k}' = k'$ (Jacobson). Hijikata and Springer have given a similar interpretation to k-forms of G_2 and F_4 (see [B]).

3. We have previously looked at the exact sequence

$$H^1(\Gamma,G) \to H^1(\Gamma,G/Z) \xrightarrow{\delta} H^2(\Gamma,Z)$$

(theorem 3.3.1). It follows from $H^1(\Gamma,G) = 1$ that δ is injective, and our classification results show that δ is, in fact, bijective.

Exercise: Determine $\bar{\Delta}$ (the restricted fundamental system) for each case in the above classification.

3.4 Classification over the real field

The following is merely a brief outline and table of classification for $k = \mathbb{R}$.[14] Cartan achieved the classification over $\mathbb{R}$ using a different point of view from the one we have taken, namely, by studying the geometry of symmetric spaces. The classification over $\mathbb{R}$ using methods similar to ours is given by Araki in Jour. of Math., Osaka City U., vol. 13, (1962).

For $k = \mathbb{R}$, Γ is of order 2; let σ_0 denote the non-identity element of Γ. For each root system $(X,\surd)$, there exists exactly one $\mathbb{R}$-compact form, which corresponds to the one-cocycle $\{\varphi_\sigma\}$ where $\varphi_{\sigma_0}{}^*: \mathcal{X} \to -\mathcal{X}$ and $\xi_{\alpha,\sigma} = \begin{cases} 1 & \text{if } \sigma = 1 \\ -1 & \text{if } \sigma = \sigma_0 \end{cases}$, and the [ ]-action of Γ is given by $[\sigma_0] = w_{\sigma_0}^{-1}\varphi_{\sigma_0}^{*-1}$. Thus w_{σ_0} is the unique element of W which maps Δ to $-\Delta$, and so $[\sigma_0]$ is the "opposition automorphism" of Δ. This opposition automorphism is well known, since it is the automorphism of Δ which when composed with w_{σ_0}, gives the automorphism $\alpha \to -\alpha$ of $(X,\surd)$. It is non-trivial only for A_ℓ, D_ℓ (ℓ odd), and E_6. Thus for A_ℓ, D_ℓ (ℓ odd) and E_6, the unique k-compact group is of type ${}^2A_\ell$, ${}^2D_\ell$, and ${}^2E_6$, respectively, and for all other simple groups, it is of type ${}^1X_\ell$.

The complete classification of simple groups of $\mathbb{R}$-rank = 1 is as follows:

${}^1A_\ell$: 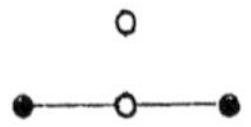 (ℓ = 1) (ℓ = 3)

${}^2A_\ell$: 

B_ℓ: 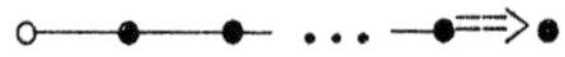

C_ℓ: 

${}^1D_\ell$: 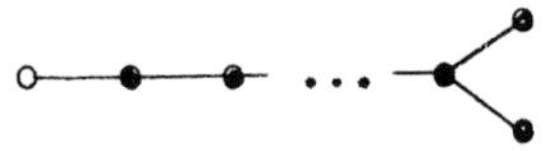 (ℓ odd)

${}^2D_\ell$:  (ℓ even)

F_4: 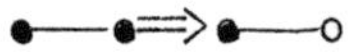

It is a remarkable contrast that in the $\mathfrak{p}$-adic case, an infinite series of diagrams occurs only for ${}^1A_\ell$, while in the real case, the infinite series occur for all other types of classical groups.

By combining these diagrams for $\mathbb{R}$-rank = 1, the following complete classification table for non-compact real forms is obtained.

Type	Γ-diagram	Cartan Notation
${}^1A_\ell$	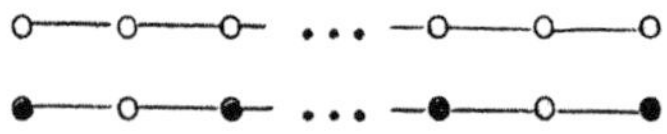	A_I A_{II}
${}^2A_\ell$	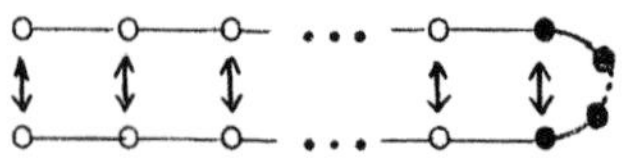	A_{III}
B_ℓ	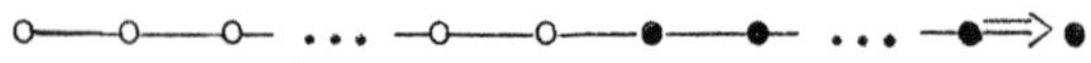	B_I
C_ℓ		C_I
	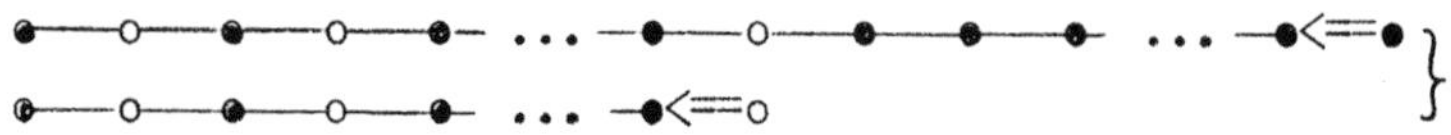	C_{II}

${}^1D_\ell$	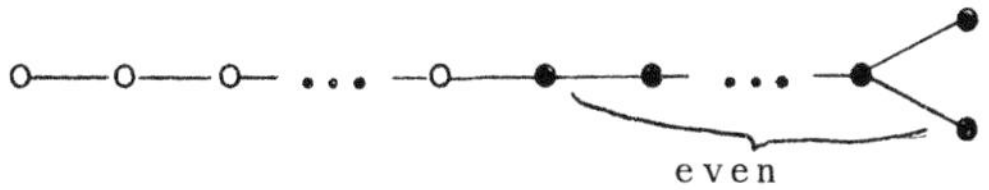	D_{I}
		D_{III}
${}^2D_\ell$	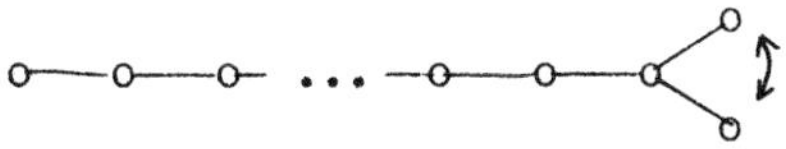	D_{I}
	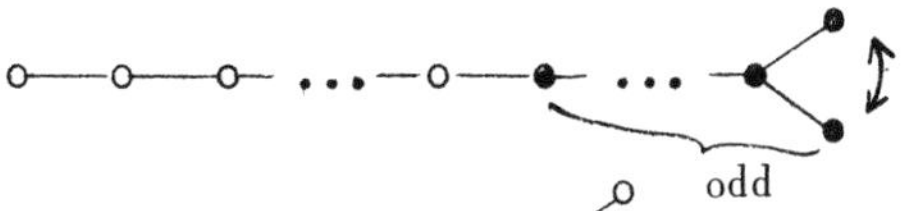	
		D_{III}
${}^1E_6$	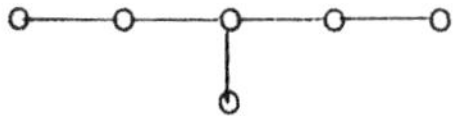	E_{I}
		E_{IV}
${}^2E_6$	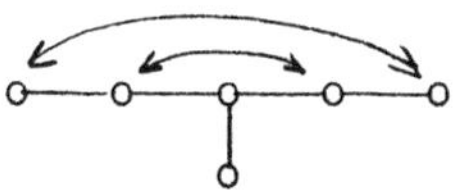	E_{II}
	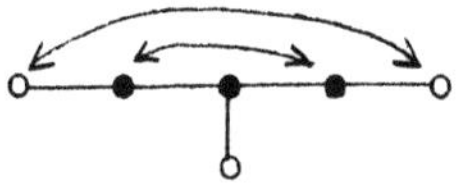	E_{III}
E_7	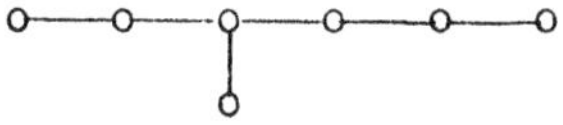	E_{V}
		E_{VII}
	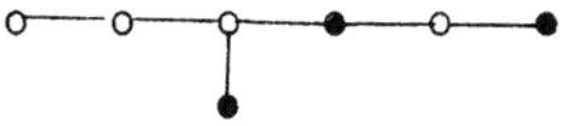	E_{VI}
E_8	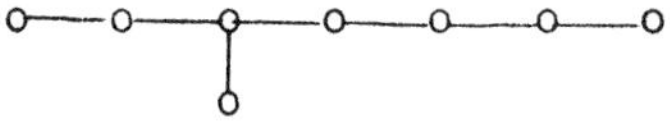	E_{VIII}
		E_{IX}
F_4	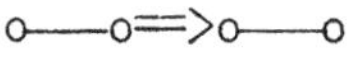	F_{I}
	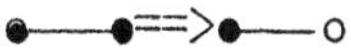	F_{II}
G_2	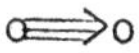	G_{I}

3.5 Open questions (algebraic number fields)

The classification for k an algebraic number field has not been completed (cf. [10]). An essential part of the problem is to determine $H^1(k, G)$ for G simply connected. If (v) denotes all valuations of k, and k_v is the completion of k with respect to v, then Borel and Serre have shown that for any algebraic group G the natural mapping

$$H^1(k, G) \to \prod_{(v)} H^1(k_v, G)$$

has finite fibers (i.e. the inverse image of any element is finite). For the case of simply connected semi-simple algebraic group G, Kneser has conjectured that this mapping is bijective; in view of Th. 3.3.1, this is equivalent to saying that the map $H^1(k, G) \to \prod_{v:\text{real}} H^1(k_v, G)$ is bijective. He has shown that for classical groups (except D_4) this amounts essentially to the classical Hasse principle for various forms, and a similar interpretation is also possible for the groups of type G_2 and F_4 (see p.122). For all other types of exceptional groups except for type E_8, Harder has verified Kneser's conjecture. Ono has clarified the relation to the Tamagawa numbers.

The following is a list of references dealing with these questions:

1) A. Borel and J.-P. Serre, "Théorème de finitude en cohomologie galoisienne", Comm. Math. Helv. 39 (1964), 111-164.

2) G. Harder, "Über die Galoiskohomologie halbeinfacher Matrizengruppen", I, Math. Z. 90 (1965), 404-428; II, ibid. 92 (1966), 396-415.

3) M. Kneser, "Hasse principle for H^1 of simply connected groups", Proc. of Symposia in Pure Math., Vol. IX (1966), 159-163.

4) T. Ono, "On the relative theory of Tamagawa numbers", Ann. of Math. 82 (1965), 88-111.

5) B.J. Veisfeiler, "Classification of semisimple Lie algebras over a $\mathfrak{p}$-adic field", Dokl. Akad. Nauk SSSR 158 (1964), 258-260 = Soviet. Math. Dokl. 5 (1964), 1206-1208.

Bibliography

[1] Borel, A., "Groupes linéaires algébriques", Annals of Math. 64 (1956), 20-82.

[2] Chevalley, C., "Classification des groupes de Lie algébriques", Séminaire C. Chevalley 1956-58.

[3] Godement, R., "Groupes linéaires algebriques sur un corps parfait", Séminaire Bourbaki 1960/61, n° 206.

[4] Lang, S., "Introduction to algebraic geometry", Interscience, 1958.

[5] Ono, T., "Arithmetic of algebraic tori", Annals of Math. 74 (1961), 101-139.

[6] Rosenlicht, M., "Some rationality questions on algebraic groups", Annali di Mat. 43 (1957), 25-50.

[7] Rosenlicht, M., "Questions of rationality for solvable algebraic groups over non-perfect fields", Annali di Mat. 61 (1963), 97-120.

[8] Satake, I., "On the theory of reductive algebraic groups over a perfect field", Jour. Math. Soc. Japan 15 (1963), 210-235.

[9] Borel, A. and Tits, J., "Groupes réductifs", I.H.E.S., Publ. de Math. n° 27 (1965), 55-151.

[10] Tits, J., "Classification of algebraic semisimple groups", [A], 33-62.

[11] Satake, I., "Symplectic representations of algebraic groups satisfying a certain analyticity condition", Acta Math. 117 (1967), 215-279.

[12] Borel, A. and Springer, T.A., "Rationality properties of linear algebraic groups II", Tôhoku Math. J. 20 (1968), 443-497.

[13] Borel, A., "Linear algebraic groups", Benjamin, 1969.

We also refer occasionally to the following books:

[A] "Algebraic Groups and Discontinuous Subgroups", Proceedings of Symposia in Pure Mathematics, vol. IX, Amer. Math. Soc., 1966.

[B] "Colloque sur la théorie des groupes algébriques", CBRM, Bruxelle, 1962.

[C] Serre, J.-P., "Corps locaux", Actualités Sci. et Ind., n° 1296, Hermann, 1962.

[D] Serre, J.-P., "Cohomologie galoisienne", Lecture Notes in Math., No. 5, 3rd ed., Springer, 1965.

[E] Bourbaki, N., "Groupes et algèbres de Lie", Chap. IV-VI, Eléments de Mathématique, Hermann, Paris, 1969.

APPENDIX:

CLASSIFICATION OVER THE REAL FIELD

by Mitsuo Sugiura

1. For the real field $\mathbb{R}$, the problem (i) in p. 87 is solved easily as follows:

Proposition 1. Let $\Omega = \mathbb{C}$. Then every semisimple algebraic group G has an $\mathbb{R}$-compact form G_o. G_o is unique up to $\mathbb{R}$-isomorphisms.

This can be proved by using "Weyl basis". See, e.g., S. Helgason, Differential Geometry and Symmetric Spaces, Ch. III, Theorem 6.3 and Corollary 7.3.

For $k = \mathbb{R}$, $\Gamma = \mathrm{Gal}(\mathbb{C}/\mathbb{R})$ is of order 2; let σ_o denote the non-identity element of Γ. By Theorem 1.3.1, Theorem 1.3.2 in II and Proposition 1 above, there exists exactly one compact $\mathbb{R}$-form for each root system $(X,\mathfrak{r})$. This compact $\mathbb{R}$-form corresponds to the one-cocycle $\{\varphi_\sigma\}$ determined by $\varphi^*_{\sigma_o} : \chi \mapsto -\chi$ and $\xi_{\alpha,\sigma} = \begin{cases} 1 & \text{if } \sigma=1 \\ -1 & \text{if } \sigma=\sigma_o \end{cases}$, and the [ ]-action of Γ is given by $[\sigma_o] = w^{-1}_{\sigma_o}\varphi^{*-1}_{\sigma_o} = -w_{\sigma_o}$. (Cf. II, §2.4) Thus w_{σ_o} is the unique element of W which maps Δ to $-\Delta$, and so $[\sigma_o]$ is the "opposition automorphism" of Δ.

Proposition 2. Let $\mathfrak{r}$ be an irreducible root system. Then the opposition automorphism of Δ is non-trivial if and only if $\mathfrak{r}$ is either $A_\ell (\ell \geqq 2)$, D_ℓ (ℓ odd) or E_6.

Thus for $\mathfrak{r} = A_\ell$ $(\ell \geqq 2)$, D_ℓ (ℓ odd) and E_6, the unique $\mathbb{R}$-compact group is of type ${}^2A_\ell$, ${}^2D_\ell$ and ${}^2E_6$, respectively, and for all other cases, it is of type ${}^1X_\ell$.

So the classification problem of $\mathbb{R}$-form is reduced to the problem (ii) in p. 87, i.e. the determination of admissible Γ-diagrams. In the following, we shall first classify more general Γ-diagrams which correspond to root systems with an action of Γ, and then pick up all admissible ones from among them by Araki's method.

2. In II, §2, we defined the Γ-diagram of a connected semisimple algebraic group G defined over k. However a Γ-diagram can be defined in a more general setting. Let $(X, \mathfrak{r})$ be an (abstract) root system. If an action $\chi \mapsto \chi^{\sigma}$ of a group Γ on $(X, \mathfrak{r})$ is given (i.e. if a homomorphism of Γ into $\mathrm{Aut}(X, \mathfrak{r})$ is given), then the Γ-diagram $(X, \mathfrak{r}, \Delta, \Delta_0, [\sigma])$ is defined as in §2.4: Δ is a Γ-fundamental system, $\Delta_0 = \Delta \cap X_0$ where $X_0 = \{\chi \mid \sum_{\sigma \in \Gamma} \chi^{\sigma} = 0\}$ and $\sigma \to [\sigma]$ is a homomorphism of Γ into $\mathrm{Aut}(X, \mathfrak{r}, \Delta)$ satisfying $\chi^{[\sigma]} = w_{\sigma}^{-1}\chi^{\sigma}$ where w_{σ} is a unique element in W_0 such that $\Delta^{\sigma} = w_{\sigma}\Delta$.

<u>Remark</u>. In our general situation, Γ-diagrams may depend on the choice of Δ, i.e. for two Γ-fundamental systems Δ, Δ', the corresponding Γ-diagrams $(X, \Delta, \Delta_0, [\sigma])$ and $(X, \Delta', \Delta'_0, [\sigma]')$ may not be congruent in general.

In our case, an action of $\Gamma = \{1, \sigma_0\}$ on $(X, \mathfrak{r})$ is determined by a single involutive element in $\mathrm{Aut}(X, \mathfrak{r})$ (denoted sometimes by s_0) determined by the action of σ_0.

<u>Proposition 3</u>. If X is a free module of rank ℓ, Δ a fundamental system of a root system $\mathfrak{r}$ in X, Δ_0 a subset of Δ, and $[\ ]$ a homomorphism of $\Gamma = \{1, \sigma_0\}$ into $\mathrm{Aut}(X, \Delta, \Delta_0)$, then the quadruple $(X, \Delta, \Delta_0, [\sigma])$ is a Γ-diagram of some action of Γ on $(X, \mathfrak{r})$ if and only if $(X'_0, \Delta_0, \Delta_0, [\sigma]|X'_0)$ is the Γ-diagram of the action $\chi^{\sigma_0} = -\chi$ of Γ on

$(X_o', \mathcal{W}_o)$, where $X_o' = \{\Delta_o\}_{\mathbf{Z}}$ and $\mathcal{W}_o = \mathcal{W} \cap X_o'$. Moreover, when that is the case, the action of Γ on $(X, \mathcal{W})$ is completely determined by the corresponding Γ-diagram $(X, \Delta, \Delta_o, [\sigma])$.

Proof. The "only if" part of Proposition 3 is obvious. Now assume that $(X_o', \Delta_o, \Delta_o, [\sigma]|X_o')$ satisfies the above condition and w_{σ_o} is the unique element in $W_o = W(\Delta_o)$ (identified with a subgroup of W) satisfying $w_{\sigma_o}\Delta_o = -\Delta_o$. Using this w_{σ_o} and $[\sigma_o]$, we define an action of Γ on X by

$$\chi^{\sigma_o} = w_{\sigma_o}\chi^{[\sigma_o]}.$$

Since both w_{σ_o} and $[\sigma_o]$ leave invariant $X_{o\mathbb{Q}}'$ and $X_{o\mathbb{Q}}'^{\perp}$ in $X_{\mathbb{Q}}$, they commute with each other. Therefore we have $\chi^{\sigma_o^2} = \chi$. So $\chi \mapsto \chi^{\sigma_o}$ defines an action of Γ on X. Moreover $(X, \Delta, \Delta_o, [\sigma])$ is the Γ-diagram of this action of Γ as is easily seen. Since the definition of χ^{σ_o} given above is the only possible way to define an action of Γ on $(X, \mathcal{W})$ giving rise to the given Γ-diagram, the last statement of Proposition 3 follows.

Remark. The following five conditions for a quadruple $\mathcal{S}_o = (X_o', \Delta_o, \Delta_o, [\sigma])$ are mutually equivalent.

(i) The quadruple $\mathcal{S}_o$ is the Γ-diagram of the action $\chi^{\sigma_o} = -\chi$ of Γ on $(X_o', \mathcal{W}_o)$.

(ii) $[\sigma_o]$ is the opposition automorphism of Δ_o.

(iii) $-[\sigma_o] \in W_o$.

(iv) The quadruple $\mathcal{S}_o$ is the Γ-diagram of an $\mathbb{R}$-compact group.

(v) If Δ_i is an irreducible component of Δ_o, then $[\sigma_o]$ leaves Δ_i invariant and $(\{\Delta_i\}_{\mathbf{Z}}, \Delta_i, \Delta_i, [\sigma]|\{\Delta_i\}_{\mathbf{Z}})$ is the Γ-diagram of one of the compact groups of type ${}^2A_\ell(\ell \geqq 2)$, ${}^2D_\ell$ (ℓ: odd), ${}^2E_6$ and ${}^1X_\ell$ where $X_\ell \neq A_\ell$ ($\ell \geqq 2$), D_ℓ (ℓ: odd) and E_6.

<u>Definition</u>. The <u>restricted rank</u> of a system $(X, \Delta, \Delta_o, [\sigma])$ is the number of orbits in $\Delta - \Delta_o$ under the action $[\ \ ]$ of the group Γ.

<u>Proposition 4</u>. There exist twenty types of Γ-irreducible Γ-diagram $\mathcal{S} = (X, \Delta, \Delta_o, [\sigma])$ of an action of Γ with the restricted rank 1. (See Table 1.)

<u>Proof</u>. First we prove that the only absolutely reducible, Γ-irreducible $\mathcal{S}$ with the restricted rank 1 is the diagram No. 1 in Table 1. In fact, every Γ-irreducible component of $\mathcal{S}_o = (X_o', \Delta_o, \Delta_o, [\sigma]|X_o')$ is absolutely irreducible, so $\Delta_o = \phi$ in this case. In the following, we assume $\mathcal{S}$ is absolutely irreducible. We give the proof for the root system A_ℓ. The proof for other types of root system can be carried out similarly. Let $\mathfrak{V} = A_\ell$. There are two possibilities. (a) $[\sigma_o] = 1$ and (b) $[\sigma_o] \neq 1$. In the case (a), a Γ-diagram $\mathcal{S} = (X, \Delta, \Delta_o, [\sigma])$ with the restricted rank 1 has the following form:

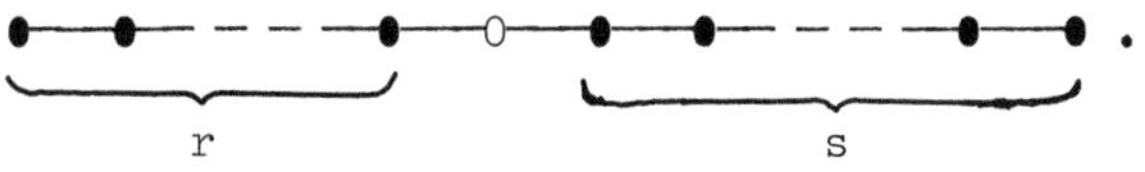

We have $\mathfrak{V}_o = A_r \times A_s$. By Proposition 3 and the above remark, $\mathcal{S}$ corresponds to an action of Γ if and only if $r \leqq 1$ and $s \leqq 1$. So the following three cases are possible:

(1) $r = s = 0$, (2) $r = 0$, $s = 1$ or $r = 1$, $s = 0$, (3) $r = s = 1$.

The Γ-diagrams corresponding to these three cases are the diagrams Nos. 2, 3 and 4 in Table 1. In the case (b), $\ell = \operatorname{rank} \mathfrak{V} \geqq 2$ and $[\sigma_o]$ is the unique element in $\operatorname{Aut}(X, \mathfrak{V}, \Delta)$ with the order 2. Hence the Γ-diagram $\mathcal{S}$ has the following form:

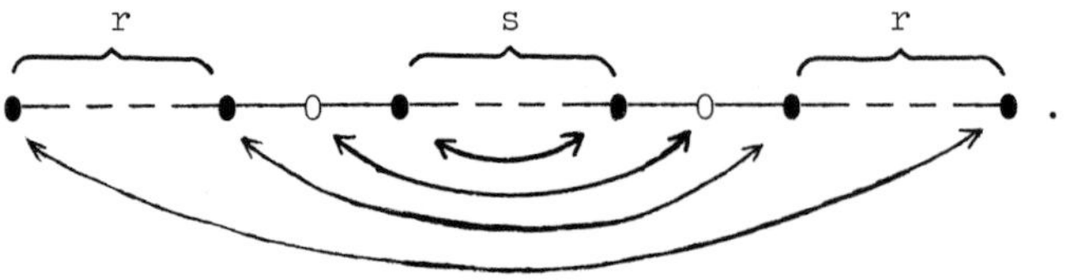

We have $\mathfrak{r}_o = A_r \times A_s \times A_r$. By Proposition 2, the Γ-diagram of the ℝ-compact form of $A_r \times A_s \times A_r$ is as follows:

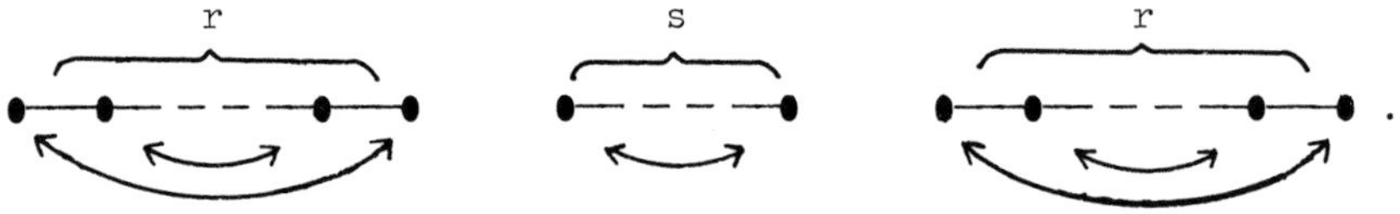

Hence by Proposition 3, the Γ-diagram $\mathcal{S}$ corresponds to an action of Γ if and only if $r = 0$. So the Γ-diagram $\mathcal{S}$ is the diagram No. 5 in Table 1.

3. Let $\underline{G} = G(X, \mathfrak{r})$ be a Chevalley group (defined over ℝ) with the root system $(X, \mathfrak{r})$ with respect to an ℝ-trivial maximal torus $\underline{T}$. Supposing an action of Γ on $(X, \mathfrak{r})$ is given, we now consider the question of its extensibility to an automorphism φ_{σ_o} of $(\underline{G}, \underline{T})$ satisfying the cocycle condition

$$(1) \qquad \varphi_{\sigma_o}^{\sigma_o} \circ \varphi_{\sigma_o} = 1.$$

(Note that, if we denote by the same symbol φ_{σ_o} the corresponding automorphism of the Lie algebra $\underline{\mathfrak{g}}$ of $\underline{G}$, then the condition (1) is equivalent to saying that the map $x \mapsto \varphi_{\sigma_o}(\bar{x})$ $(x \in \underline{\mathfrak{g}})$ is an "anti-involution" of $\underline{\mathfrak{g}}$.) If such an extension exists, then one has in the notation of p. 89

$$(2) \qquad \varphi_{\sigma_o} \longleftrightarrow \{\varphi_{\sigma_o}^*, \mu_\alpha \ (\alpha \in \mathfrak{r})\},$$

where $\varphi_{\sigma_o}^* \in \mathrm{Aut}(X, \mathfrak{r})$ should coincide with the given action of σ_o on

$(X, \mathfrak{W})$. To simplify the notation, we denote in the following this action of σ_o by bar. Hence one has

(3) $$\varphi^*_{\sigma_o}(\chi) = \bar{\chi} \qquad \text{for} \quad \chi \in X.$$

We further assume that the system $\{\mu_\alpha\}$ is defined by a (fixed) Weyl basis $\{E_\alpha\}$ of the Lie algebra $\underline{\mathfrak{g}}$, i.e. one has

(4) $$\varphi_{\sigma_o}(E_\alpha) = \mu_\alpha E_{\bar{\alpha}} \qquad \text{for} \quad \alpha \in \mathfrak{W}.$$

Then, one has

(5) $$\begin{cases} \mu_{\alpha+\beta} = \dfrac{N_{\alpha,\beta}}{N_{\bar{\alpha},\bar{\beta}}} \mu_\alpha \cdot \mu_\beta & \text{if} \quad \alpha,\ \beta,\ \alpha+\beta \in \mathfrak{W}, \\ \mu_{-\alpha} = \mu_\alpha^{-1}\ . \end{cases}$$

Therefore $\{\mu_\alpha\}$ is uniquely determined by $\{\mu_{\alpha_i}\ (\alpha_i \in \Delta)\}$. Conversely, it is known (Lemma 4, p. 92) that any given ℓ-tuple (μ_{α_i}) in $\mathbb{C}^{*\ell}$ can be extended uniquely to a system $\{\mu_\alpha\ (\alpha \in \mathfrak{W})\}$ satisfying (5), so that it determines $\varphi_{\sigma_o} \in \mathrm{Aut}(\underline{G}, \underline{T})$ by the relation (2). The condition (1) on φ_{σ_o} is clearly equivalent to

(6) $$\bar{\mu}_\alpha \mu_{\bar{\alpha}} = 1 \qquad \text{for} \quad \alpha \in \mathfrak{W}.$$

Thus, the action of Γ on $(X, \mathfrak{W})$ is extendible to an automorphism φ_{σ_o} satisfying (1), if and only if one can find an ℓ-tuple $(\mu_{\alpha_i})_{1\le i\le \ell}$ such that the condition (6) is satisfied for its unique extension $\{\mu_\alpha (\alpha \in \mathfrak{W})\}$.

Now, the condition $N_{-\alpha,-\beta} = -N_{\alpha,\beta}$ for a Weyl basis implies also $|N_{\bar{\alpha},\bar{\beta}}| = |N_{\alpha,\beta}|$. Hence, for a system $\{\mu_\alpha\}$ satisfying (5), we may always assume, by replacing $\{\mu_\alpha\}$ by $\{\mu_\alpha/|\mu_\alpha|\}$, that

(7) $$|\mu_\alpha| = 1 \qquad \text{for} \quad \alpha \in \mathfrak{r},$$

(without changing the cohomology class of $\{1, \varphi_{\sigma_0}\}$, when it satisfies (6)). Then the relations (5), (6), (7) imply

$$\mu_\alpha = \pm 1 \qquad \text{for} \quad \alpha \in \mathfrak{r}_0.$$

Let (G, T) be the $\mathbb{C}/\mathbb{R}$-form of $(\underline{G}, \underline{T})$ corresponding to the cocycle $\{1, \varphi_{\sigma_0}\}$ and A the subtorus of T corresponding to X_0. Then A is a maximal $\mathbb{R}$-trivial torus in G if and only if $G(\mathfrak{r}_0)$ is $\mathbb{R}$-compact, and, by Proposition 1, this occurs if and only if one has

(8) $$\mu_\alpha = -1 \qquad \text{for all} \quad \alpha \in \mathfrak{r}_0.$$

(See also Araki [14], Prop. 1.1. Note that our normalization of Weyl basis is different from that in [14].) It is clear that the condition (6) and (8) are equivalent to

(6') $$\bar{\mu}_{\alpha_i}\mu_{\bar{\alpha}_i} = 1 \qquad \text{for} \quad \alpha_i \in \Delta - \Delta_0,$$

(8') $$\mu_{\alpha_i} = -1 \qquad \text{for} \quad \alpha_i \in \Delta_0.$$

Thus, summing up, one obtains the following criterion.

<u>Proposition 5</u>. Let $\mathcal{S} = (X, \Delta, \Delta_0, [\sigma])$ be a Γ-diagram of a root system with an action of Γ (denoted by bar). Then $\mathcal{S}$ is admissible, if and only if one can find an ℓ-tuple $(\mu_{\alpha_i})_{1\le i\le \ell}$ with $|\mu_{\alpha_i}| = 1$ such that the conditions (6') and (8') are satisfied for its unique extension $\{\mu_\alpha (\alpha \in \mathfrak{r})\}$.

4. <u>Definition</u>. A root system $(X, \mathfrak{r})$ with an action of $\Gamma = \{1, \sigma_0\}$ (denoted by bar) is said to be <u>normal</u> if $\bar{\alpha} - \alpha \notin \mathfrak{r}$ for any root α in $\mathfrak{r}$.

The corresponding Γ-diagram is also called normal.

Proposition 6 (Araki). If a root system $(X, \mathfrak{W})$ with an action of Γ is admissible (i.e. corresponds to (G, T)), then it is normal.

Proof. Suppose the system is admissible, i.e. the action of Γ on $(X, \mathfrak{W})$ is extended to an automorphism $\varphi_{\sigma_o} \leftrightarrow \{(\text{bar}), \mu_\alpha\ (\alpha \in \mathfrak{W})\}$ satisfying (1), (7), (8). If $\bar{\alpha} = -\alpha$, then $\bar{\alpha} - \alpha = -2\alpha \notin \mathfrak{W}$. Now suppose that $\bar{\alpha} \neq -\alpha$ and let $\beta = \bar{\alpha} - \alpha \in \mathfrak{W}$. Then β belongs to $\mathfrak{W}_o$, so that one has $\varphi_{\sigma_o}(E_\beta) = -E_{-\beta}$. Applying $\varphi_{\sigma_o}^{\sigma_o}$ on the both sides of $[E_\alpha, \varphi_{\sigma_o}(E_{-\alpha})] = \mu_{-\alpha} N_{\alpha,-\bar{\alpha}} E_{-\beta}$, we have

$$[\varphi_{\sigma_o}^{\sigma_o}(E_\alpha), E_{-\alpha}] = -\mu_{-\alpha} N_{\alpha,-\bar{\alpha}} E_\beta.$$

On the other hand, we have

$$[\varphi_{\sigma_o}^{\sigma_o}(E_\alpha), E_{-\alpha}] = \bar{\mu}_\alpha N_{\bar{\alpha},-\alpha} E_\beta .$$

Hence we have

$$-\mu_{-\alpha} N_{\alpha,-\bar{\alpha}} = \bar{\mu}_\alpha N_{\bar{\alpha},-\alpha} = \bar{\mu}_\alpha N_{\alpha,-\bar{\alpha}} ,$$

and $-\mu_{-\alpha} = \bar{\mu}_\alpha$. Since $\mu_\alpha \mu_{-\alpha} = 1$, the last equality implies $\mu_\alpha \bar{\mu}_\alpha = -1$. This contradition proves Proposition 6.

Proposition 7. Among the Γ-diagrams listed in Table 1, the diagrams Nos. 3,7,8,17,19,20 are not normal. All other diagrams in Table 1 are normal.

Proof. For a non-normal system, we give the action of σ_o as an element of $\mathrm{Aut}(X, \mathfrak{W})$, denoted by s_o, and the root α such that $\bar{\alpha} - \alpha \in \mathfrak{W}$. We use the notation of roots in Bourbaki [E].

No. 3. $s_o = s_{\varepsilon_2 - \varepsilon_3}$, $\alpha = \varepsilon_1 - \varepsilon_2$, $\bar{\alpha} - \alpha = \varepsilon_2 - \varepsilon_3$.

No. 7. $s_o = s_{\varepsilon_1-\varepsilon_2} s_{\varepsilon_3} \cdots s_{\varepsilon_\ell}$, $\alpha = \varepsilon_2$, $\bar{\alpha} - \alpha = \varepsilon_1 - \varepsilon_2$.

No. 8. $s_o = s_{2\varepsilon_2} \cdots s_{2\varepsilon_\ell}$, $\alpha = \varepsilon_1 + \varepsilon_2$, $\bar{\alpha} - \alpha = -2\varepsilon_2$.

No. 17. $s_o = s_{\varepsilon_1-\varepsilon_2} s_{\varepsilon_3} s_{\varepsilon_4}$, $\alpha = \varepsilon_1$, $\bar{\alpha} - \alpha = \varepsilon_2 - \varepsilon_1$.

No. 19. $s_o = s_{\varepsilon_1-\varepsilon_2}$, $\alpha = \varepsilon_1 - \varepsilon_3$, $\bar{\alpha} - \alpha = \varepsilon_2 - \varepsilon_1$.

No. 20. $s_o = s_{-2\varepsilon_1+\varepsilon_2+\varepsilon_3}$, $\alpha = \varepsilon_1 - \varepsilon_2$, $\bar{\alpha} - \alpha = -2\varepsilon_1 + \varepsilon_2 + \varepsilon_3$.

To prove a system is normal, it is sufficient to show $\bar{\alpha} - \alpha \notin \mathfrak{r}$ for all $\alpha \in \mathfrak{r} - \mathfrak{r}_o$.

No. 11 and No. 13. In these cases, one has $X_{o_{\mathbb{Q}}} = \mathbb{Q}(\varepsilon_1-\varepsilon_2) + \sum_{i=3}^{\ell} \mathbb{Q}\varepsilon_i$. So we have

$$\mathfrak{r}^+ - \mathfrak{r}_o = \{\varepsilon_1 + \varepsilon_2,\ \varepsilon_1 \pm \varepsilon_i,\ \varepsilon_2 \pm \varepsilon_i\ (i \geqq 3)\}.$$

Since $\overline{\varepsilon_1+\varepsilon_2} - (\varepsilon_1+\varepsilon_2) = 0 \notin \mathfrak{r}$, $\overline{\varepsilon_1\pm\varepsilon_i} - (\varepsilon_1\pm\varepsilon_i) = \varepsilon_2 - \varepsilon_1 \pm 2\varepsilon_i \notin \mathfrak{r}$ and $\overline{\varepsilon_2\pm\varepsilon_i} - (\varepsilon_2\pm\varepsilon_i) = \varepsilon_1 - \varepsilon_2 + 2\varepsilon_i \notin \mathfrak{r}$, the diagrams Nos. 11 and 13 are normal. Similarly we can prove that the diagrams Nos. 1,2,4,5,6,9,10,12,14,15,16 and 18 are normal.

Now we determine admissible systems with $\mathbb{R}$-rank 1. The following identities between the structure constants $N_{\alpha,\beta}$ for the Weyl base $\{E_\alpha\}$ are well known. (Cf. Helgason, loc. cit.)

(A) If α, β, γ are roots and satisfy $\alpha + \beta + \gamma = 0$, then

$$N_{\alpha,\beta} = N_{\beta,\gamma} = N_{\gamma,\alpha}.$$

(B) If $\alpha, \beta, \gamma, \delta$ are roots and satisfy $\alpha + \beta + \gamma + \delta = 0$, $\beta + \gamma \neq 0$, $\gamma + \delta \neq 0$ and $\delta + \beta \neq 0$, then

$$N_{\alpha,\beta}N_{\gamma,\delta} + N_{\alpha,\gamma}N_{\delta,\beta} + N_{\alpha,\delta}N_{\beta,\gamma} = 0.$$

Proposition 8 (Araki). Let $(X, \mathfrak{r})$ be a root system with an action of Γ (denoted by bar), and let $\{\mu_\alpha\ (\alpha \in \mathfrak{r})\}$ be a system satisfying (5) and (7). Then we have the following results.

1) If α is a root in $\mathfrak{r} - \mathfrak{r}_0$ and there exist two roots γ and δ in $\mathfrak{r}_0$ such that $\alpha + \gamma,\ \alpha + \delta \in \mathfrak{r},\ \gamma + \delta \notin \mathfrak{r} \cup \{0\}$ and $\bar{\alpha} = \alpha + \gamma + \delta$, then we have

$$\bar{\mu}_\alpha \mu_{\bar{\alpha}} = 1.$$

2) If α is a root in $\mathfrak{r} - \mathfrak{r}_0$ and there exist three roots $\gamma, \delta, \varepsilon$ in $\mathfrak{r}_0$ such that $\alpha + \gamma,\ \alpha + \delta,\ \alpha + \varepsilon,\ \bar{\alpha} - \gamma,\ \bar{\alpha} - \delta$ and $\bar{\alpha} - \varepsilon$ are roots, $\gamma + \delta,\ \gamma + \varepsilon$ and $\delta + \varepsilon$ do not belong to $\mathfrak{r} \cup \{0\}$, and $\bar{\alpha} = \alpha + \gamma + \delta + \varepsilon$, then we have

$$\bar{\mu}_\alpha \mu_{\bar{\alpha}} = -1.$$

Proof. 1) Applying φ_{σ_0} on both sides of

$$[[E_\alpha, E_\gamma], E_\delta] = N_{\alpha,\gamma} N_{\alpha+\gamma,\delta} E_{\bar{\alpha}},$$

and comparing the coefficients, we get

$$\bar{\mu}_\alpha \mu_{\bar{\alpha}} = (N_{\alpha+\gamma+\delta,-\gamma} N_{\alpha+\delta,-\delta})/(N_{\alpha,\gamma} N_{\alpha+\gamma,\delta}) \tag{9}$$

by the relations (7), (8). Applying the above identity (B) to the quadruple $\{\alpha + \gamma + \delta, -\alpha, -\gamma, -\delta\}$, we get

$$N_{\alpha+\gamma+\delta,-\gamma} N_{-\delta,-\alpha} + N_{\alpha+\gamma+\delta,-\delta} N_{-\alpha,-\gamma} = 0.$$

Therefore we have

$$N_{\alpha+\gamma+\delta,-\gamma}/N_{\alpha,\gamma} = N_{\alpha+\gamma+\delta,-\delta}/N_{\alpha,\delta}. \tag{10}$$

Next applying the identities (A) to the triples $\{\alpha+\gamma+\delta,-\delta,-\alpha-\gamma\}$ and $\{\alpha+\delta,-\delta,-\alpha\}$, we get

(11) $$N_{\alpha+\gamma+\delta,-\delta} = N_{-\delta,-\alpha-\gamma} = N_{\alpha+\gamma,\delta}$$

and

(12) $$N_{\alpha+\delta,-\delta} = N_{-\delta,-\alpha} = N_{\alpha,\delta}.$$

By the relations (9), (10), (11) and (12), we conclude that $\bar{\mu}_\alpha \mu_{\bar{\alpha}} = 1$.

2) Applying φ_{σ_o} on both sides of

$$[[[E_\alpha, E_\gamma], E_\delta], E_\varepsilon] = N_{\alpha,\gamma} N_{\alpha+\gamma,\delta} N_{\alpha+\gamma+\delta,\varepsilon} E_{\bar{\alpha}},$$

and comparing the coefficients, we get

(13) $$-\bar{\mu}_\alpha \mu_{\bar{\alpha}} = (N_{\alpha+\gamma+\delta+\varepsilon,-\gamma} N_{\alpha+\delta+\varepsilon,-\delta} N_{\alpha+\varepsilon,-\varepsilon})/(N_{\alpha,\gamma} N_{\alpha+\gamma,\delta} N_{\alpha+\gamma+\delta,\varepsilon}).$$

On the other hand, applying the identity (B) to the quadruple $\{\alpha+\gamma+\delta+\varepsilon,-\alpha-\delta,-\gamma, -\varepsilon\}$, we see that

$$N_{\alpha+\gamma+\delta+\varepsilon,-\gamma} N_{-\varepsilon,-\alpha-\delta} + N_{\alpha+\gamma+\delta+\varepsilon,-\varepsilon} N_{-\alpha-\delta,-\gamma} = 0,$$

hence

(14) $$N_{\alpha+\gamma+\delta+\varepsilon,-\gamma} / N_{\alpha+\delta,\gamma} = N_{\alpha+\gamma+\delta+\varepsilon,-\varepsilon} / N_{\alpha+\delta,\varepsilon}.$$

Next applying (A) to the triple $\{\alpha+\gamma+\delta+\varepsilon, -\varepsilon, -\alpha-\gamma-\delta\}$, we get

(15) $$N_{\alpha+\gamma+\delta+\varepsilon,-\varepsilon} = N_{-\varepsilon,-\alpha-\gamma-\delta} = N_{\alpha+\gamma+\delta,\varepsilon}.$$

By the relations (14) and (15), we have

(16) $$N_{\alpha+\gamma+\delta+\varepsilon,-\gamma} / N_{\alpha+\gamma+\delta,\varepsilon} = N_{\alpha+\delta,\gamma} / N_{\alpha+\delta,\varepsilon}.$$

Similarly, applying (B) to the quadruple $\{\alpha+\delta+\varepsilon, -\alpha, -\delta, -\varepsilon\}$ and then applying (A) to the triple $\{\alpha+\delta+\varepsilon, -\varepsilon, -\alpha-\delta\}$, we have

(17) $$N_{\alpha+\delta+\varepsilon,-\delta}/N_{\alpha,\delta} = N_{\alpha+\delta,\varepsilon}/N_{\alpha,\varepsilon} .$$

Further, applying (A) to the triple $\{\alpha+\varepsilon, -\varepsilon, -\alpha\}$, we get

(18) $$N_{\alpha+\varepsilon,-\varepsilon} = N_{-\varepsilon,-\alpha} = N_{\alpha,\varepsilon} .$$

Finally, since $[E_\gamma, E_\delta] = 0$, we have the equality

$$[[E_\alpha, E_\gamma], E_\delta] = [[E_\alpha, E_\delta], E_\gamma],$$

which implies that

(19) $$N_{\alpha,\delta}/(N_{\alpha,\gamma}N_{\alpha+\gamma,\delta}) = 1/N_{\alpha+\delta,\gamma} .$$

Multiplying (16), (17), (18) and (19) side by side, and then comparing with (13), we conclude that $\bar{\mu}_\alpha \mu_{\bar{\alpha}} = -1$.

Theorem 1. There exist exactly nine types of ℝ-irreducible admissible systems with ℝ-rank 1. They are the systems corresponding to the diagrams Nos. 1,2,4,5,6,9,10,12 and 18 in Table 1.

Proof. By Propositions 4,6 and 7, it is sufficients to prove that the diagrams Nos. 1,2,4,5,6,9,10,12 and 18 are admissible and the diagrams Nos. 11,13, 14,15,16 are not admissible. To prove this, we check the condition in Proposition 5. In the following we use the notation of roots in Bourbaki [E].

No. 1. Let $\Delta = \{\alpha_1, \alpha_2\}$ and $\mu_{\alpha_1} = \mu_{\alpha_2}$ be an arbitrary complex number with the absolute value 1. Then the condition (6') is satisfied because $\bar{\alpha}_1 = \alpha_2$, $\bar{\alpha}_2 = \alpha_1$.

No. 2. Let μ_{α_1} be an arbitrary complex number with the absolute value 1. Then one has (6') because $\bar{\alpha}_1 = \alpha_1$.

No. 5. Let μ_{α_1} be an arbitrary complex number with the absolute value 1 and put

$$\mu_{\alpha_\ell} = -\mu_{\alpha_1} N_{\alpha_\ell, \beta} / N_{\bar{\alpha}_\ell, -\beta} \quad \text{where} \quad \beta = \alpha_2 + \ldots + \alpha_{\ell-1} = \varepsilon_2 - \varepsilon_\ell .$$

Since $\bar{\alpha}_1 = \alpha_\ell + \beta$ and $\bar{\beta} = -\beta$ we get

$$-\mu_{\alpha_\ell} N_{\bar{\alpha}_\ell, -\beta} = N_{\alpha_\ell, \beta} \mu_{\bar{\alpha}_1} ,$$

by applying φ_{σ_0} on both sides of $[E_{\alpha_\ell}, E_\beta] = N_{\alpha_\ell, \beta} E_{\bar{\alpha}_1}$. So we have $\bar{\mu}_{\alpha_1} \mu_{\bar{\alpha}_1} = -\bar{\mu}_{\alpha_1} \mu_{\alpha_\ell} N_{\bar{\alpha}_\ell, -\beta} / N_{\alpha_\ell, \beta} = |\mu_{\alpha_1}|^2 = 1$. Similarly, we have $-\mu_{\alpha_1} N_{\bar{\alpha}_1, -\beta} = N_{\alpha_1, \beta} \mu_{\bar{\alpha}_\ell}$ and

$$\bar{\mu}_{\alpha_\ell} \mu_{\bar{\alpha}_\ell} = \bar{\mu}_{\alpha_1} \mu_{\bar{\alpha}_1} N_{\alpha_\ell, \beta} N^{-1}_{\bar{\alpha}_\ell, -\beta} N_{\bar{\alpha}_1, -\beta} N^{-1}_{\alpha_1, \beta} = 1,$$

because $N_{\bar{\alpha}_\ell, -\beta} = N_{-\beta, -\alpha_1} = N_{\alpha_1, \beta}$ and $N_{\bar{\alpha}_1, -\beta} = N_{-\beta, -\alpha_\ell} = N_{\alpha_\ell, \beta}$.

The fact that the diagrams Nos. 4,6,9,10,12 and 18 in Table 1 satisfy the condition in Proposition 5 is proved by Proposition 8,1). In the following, we give the action s_0 of $\sigma_0 \in \Gamma$ on $(X, \mathfrak{W})$ or X_{0_Q} and the roots α, γ and δ satisfying the condition in Proposition 8,1).

No. 4. $s_0 = s_{\varepsilon_1 - \varepsilon_2} s_{\varepsilon_3 - \varepsilon_4}$, $\alpha = \alpha_2 = \varepsilon_2 - \varepsilon_3$, $\gamma = \alpha_1 = \varepsilon_1 - \varepsilon_2$, $\delta = \alpha_3 = \varepsilon_3 - \varepsilon_4$.

No. 6. $s_0 = s_{\varepsilon_2} s_{\varepsilon_3} \cdots s_{\varepsilon_\ell}$, $\alpha = \alpha_1 = \varepsilon_1 - \varepsilon_2$, $\gamma = \delta = \varepsilon_2$.

No. 9. $s_0 = s_{\varepsilon_1 - \varepsilon_2} s_{2\varepsilon_2} \cdots s_{2\varepsilon_\ell}$, $\alpha = \alpha_2 = \varepsilon_2 - \varepsilon_3$, $\gamma = \alpha_1 = \varepsilon_1 - \varepsilon_2$, $\delta = 2\varepsilon_3$.

No. 10 and 12. $X_{0_Q} = \sum_{i=2}^{\ell} Q\varepsilon_i$, $\alpha = \alpha_1 = \varepsilon_1 - \varepsilon_2$, $\gamma = \varepsilon_2 - \varepsilon_\ell$, $\delta = \varepsilon_2 + \varepsilon_\ell$.

No. 18. $s_0 = s_{\varepsilon_2 - \varepsilon_3} s_{\varepsilon_3 - \varepsilon_4} s_{\varepsilon_4}$, $\alpha = \alpha_4 = 2^{-1}(\varepsilon_1 - \varepsilon_2 - \varepsilon_3 - \varepsilon_4)$, $\gamma = \varepsilon_2$, $\delta = \varepsilon_3 + \varepsilon_4$.

The fact that the diagrams Nos. 11, 13,14,15,16 and 17 do not satisfy the condition in Proposition 5 is proved by Proposition 8,2). We give the action

s_o of σ_o on $(X, \mathcal{W})$ or $X_{o_{\mathbb{Q}}}$ and the roots α, γ, δ and ε which satisfy the conditions of Proposition 8,2).

No. 11 and 13. $X_{o_{\mathbb{Q}}} = \mathbb{Q}(\varepsilon_1 - \varepsilon_2) + \sum_{i=3}^{\ell} \mathbb{Q}\varepsilon_i$, $\alpha = \alpha_2 = \varepsilon_2 - \varepsilon_3$,

$\gamma = \alpha_1 = \varepsilon_1 - \varepsilon_2$, $\delta = \varepsilon_3 - \varepsilon_\ell$, $\varepsilon = \varepsilon_3 + \varepsilon_\ell$.

No. 14. $X_{o_{\mathbb{Q}}} = \{\mathbb{Q}(\varepsilon_1+\varepsilon_2+\varepsilon_3+\varepsilon_4+\varepsilon_5-\varepsilon_6-\varepsilon_7+\varepsilon_8)\}^{\perp}$, $\alpha = \alpha_2$,

$\gamma = \alpha_1 + \alpha_3 + \alpha_4$, $\delta = \alpha_3 + \alpha_4 + \alpha_5$, $\varepsilon = \alpha_4 + \alpha_5 + \alpha_6$.

No. 15. $X_{o_{\mathbb{Q}}} = \{\mathbb{Q}(\varepsilon_7 - \varepsilon_8)\}^{\perp}$, $\alpha = \alpha_1$, $\gamma = \alpha_2 + \alpha_3 + 2\alpha_4 + \alpha_5 + \alpha_6 = \varepsilon_3 + \varepsilon_4$,

$\delta = \alpha_2 + \alpha_3 + \alpha_4 + \alpha_5 + \alpha_6 = \varepsilon_2 + \varepsilon_5$,

$\varepsilon = \alpha_3 + \alpha_4 + \alpha_5 + \alpha_6 + \alpha_7 = \varepsilon_6 - \varepsilon_1$.

No. 16. $X_{o_{\mathbb{Q}}} = \{\mathbb{Q}(\varepsilon_7 + \varepsilon_8)\}^{\perp}$, $\alpha = \alpha_8 = \varepsilon_7 - \varepsilon_6$,

$\gamma = \alpha_3 + \alpha_4 + \alpha_5 + \alpha_6 + \alpha_7 = \varepsilon_6 - \varepsilon_1$,

$\delta = \alpha_2 + \alpha_4 + \alpha_5 + \alpha_6 + \alpha_7 = \varepsilon_6 + \varepsilon_1$,

$\varepsilon = 2\alpha_1 + 2\alpha_2 + 3\alpha_3 + 4\alpha_4 + 3\alpha_5 + 2\alpha_6 + \alpha_7 = \varepsilon_8 - \varepsilon_7$.

Thus we have proved that the admissible $\mathbb{R}$-irreducible systems with $\mathbb{R}$-rank 1 are exhausted by the diagrams Nos. 1,2,4,5,6,9,10,12 and 18 in Table 1. It is clear that any two of them are not congruent. q.e.d.

Theorem 1 enables us to determine the admissible $\mathbb{R}$-irreducible systems by using Proposition 3.1.2 in II. An $\mathbb{R}$-irreducible and non absolutely irreducible system corresponds to a group of type $R_{\mathbb{C}/\mathbb{R}}(G)$ where G is a simple algebraic group defined over $\mathbb{C}$. So we can restrict ourselves to the absolutely irreducible case.

Theorem 2. The $\mathbb{R}$-irreducible and absolutely irreducible systems $\mathcal{S} = (X, \Delta, \Delta_o, [\sigma])$ which are not $\mathbb{R}$-compact are exhausted by the systems in Table on p. 124-125.

The proof is easily obtained by Theorem 1 and Proposition 3.1.2 in II.

Table 1

No.	Type	Γ-diagram	normal	admissible
1	$A_1 \times A_1$		+	+
2	A_1		+	+
3	A_2		–	–
4	A_3	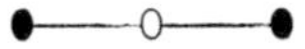	+	+
5	A_ℓ		+	+
6	B_ℓ		+	+
7			–	–
8	C_ℓ		–	–
9			+	+
10	D_{2n}		+	+
11	$(n \geqq 2)$		+	–
12	D_{2n+1}		+	+
13	$(n \geqq 2)$		+	–
14	E_6	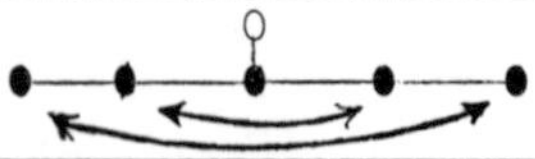	+	–
15	E_7		+	–
16	E_8	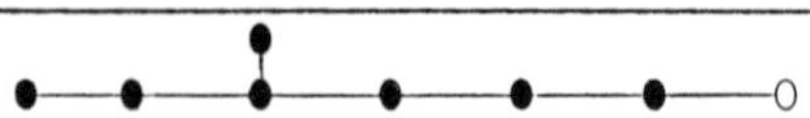	+	–
17	F_4		–	–
18			+	+
19	G_2		–	–
20			–	–

References

[14] S. Araki, On root systems and an infinitesimal classification of irreducible symmetric spaces, Journal of Math., Osaka City University 13(1962), 1-34.

[E] N. Bourbaki, Groupes et algèbres de Lie, Ch. IV, V, VI, Paris (1968), Hermann.

FOOT NOTES

P.10, 1) Two connected algebraic groups G and G' are said to be <u>isogeneous</u> if there exist a connected algebraic group G'' and isogenies $\varphi : G'' \to G$, $\varphi' : G'' \to G'$. It is easy to see that this relation is an equivalence relation.

P.16, 2) In the following, p denotes the "characteristic exponent", i.e. if p_0 is the characteristic of the universal domain, one puts $p = p_0$ if $p_0 \neq 0$ and $p = 1$ if $p_0 = 0$.

P.24, 3) Precisely speaking, $\mathfrak{A}_k$ (the set of k-rational points of $\mathfrak{A}$) is a normal simple algebra over k in the usual sense. $\mathfrak{A}$ itself is an algebra over the universal domain having a "k-structure" in the sense of p.12.

P.33, 4) In these Notes, a connected algebraic group G is called <u>simple</u> if G is semi-simple, of dimension > 0 and does not have any proper closed normal subgroup of dimension > 0. Thus, G may have a non-trivial center which is a finite abelian group. (A simple group in this sense is sometimes called "almost simple".)

P.46, 5) See I. Satake, "On a theorem of E. Cartan", J. Math. Soc. Japan 2 (1951), 284-305.

P.50, 6) For the classification of Dynkin diagrams, see van der Waerden, Math. Z. 37 (1933), E. Witt, Abh. Math. Sem. Hamburg Univ. 14 (1941), and Bourbaki [E], Ch. VI, §4.

P.54, 7) For Chevalley groups, see A. Borel, "Properties and linear representations of Chevalley groups", Seminar on Algebraic Groups and Related Finite Groups, Lecture-Notes in Math. 131, Springer-Verlag, 1970. Cf. also, R. Steinberg, "Lectures on Chevalley groups", Yale Univ., 1967.

P.54, 7a) By definition, the Chevalley group $G(X, \mathcal{W})$ is a connected semi-simple algebraic group defined over the prime field k_0, having a maximal torus T defined and trivial over k_0, such that its root system with respect to T is identified with $(X, \mathcal{W})$. Later on, the word "Chevalley group" will sometimes be used as an abbreviation of "group of Chevalley type".

P.58, 8) More precisely, one has $\Theta \cong \mathrm{Aut}(X, \Delta)$ (Rem. 2, p.59) and, when G is a Chevalley group over the prime field k_0, Θ can be chosen in such a way that all elements $\theta \in \Theta$ are k_0-rational (Rem., p.92).

P.60, 9) The statement of this Proposition (along with the following proof) remains valid, if one replaces the prime field k_0 by any perfect field k (cf. Coroll. 2.4.2. (1), p.97).

P.74, 10) In the characteristic zero case, this decomposition can easily be obtained from the corresponding decomposition of the Lie algebra. For the general case, see [3], [9].

P.84, 10a) The main idea of the classification as explained here was first announced by J. Tits (C. R. Acad. Sci. Paris 249 (1959), 1438-1440; see also [10]). Beyond the Witt type theorem (Th. 2.4.1, p.87) he has also given various necessary (and sufficient) conditions for a solution of the problem (ii) (p.87). Here we do not discuss these conditions, for we can dispense with them in the case of local fields by taking a different approach.

P.102, 11) Every admissible Γ-diagram $\mathcal{S}$ can be "built up" (in the sense of Prop. 3.1.2.) from the admissible Γ-diagrams of k-rank 1 as follows. Let $\mathcal{S} = (X, \Delta, \Delta_0, [\sigma])$ be the Γ-diagram of (G, T) and let $\overline{\Delta} = \pi(\Delta - \Delta_0) = \{\gamma_1, \ldots, \gamma_r\}$. For each γ_i, define $\Delta^i = \Delta_0 \cup (\pi^{-1}(\gamma_i) \cap \Delta)$. Then Δ^i satisfies the conditions of Lemma 3.1.1, (since Δ_0 is $[\sigma]$-invariant and $\pi(\alpha) = \pi(\alpha^{[\sigma]}) = \gamma_i$ for all $\alpha \in \Delta^i - \Delta_0$). Therefore, the canonical subsystem $\mathcal{S}_{\Delta^i}$ is admissible, and is the Γ-diagram of the subgroup $G(\Delta^i)$ of k-rank 1. The decomposition $\Delta = \Delta^1 \cup \ldots \cup \Delta^r$ clearly satisfies the conditions of Prop. 3.1.2.

P.103, 12) In classifying the absolutely irreducible Γ-diagrams of k-rank 1, it would be useful to observe that such a diagram $(X, \Delta, \Delta_0, [\sigma])$ should satisfy the following "necessary conditions": (i) Δ_0 is invariant under the opposition automorphism, (ii) $\Delta - \Delta_0$ is a [Γ]-orbit. For the condition (i), see Tits' paper in [A], or Schattschneider's paper quoted at p.69.

P.110, 13) The approach adopted in this section is essentially due to M. Kneser, (Math. Z. 89 (1965)).

P.123, 14) A more detailed account is given in the Appendix by Sugiura, where Araki's method is reproduced in a simplified form, along with a more general classification of root systems with an involution. For another approach to the classification over the real, see S. Murakami, Osaka J. Math. 2 (1965).

LIST OF NOTATION

INDEX OF TERMS

about the book . . .

The purpose of this work is to explain the general principle of classification theory of semi-simple algebraic groups. A few examples have been presented where the classification is known. To make the text self-contained, the basic theory on algebraic groups is also outlined in the first chapter. The material is based on lecture notes given at the University of Chicago in 1967. In addition, an appendix written by M. Sugiura of the University of Tokyo has been included; its new efficient way of classifying real simple algebraic groups simplifies previous methods.

The book is directed toward graduate students working in algebraic groups and related areas.

about the author . . .

DR. ICHIRO SATAKE is Professor in the Department of Mathematics at the University of California in Berkeley; his prime research interests are algebraic groups and automorphic functions. Dr. Satake received his B.Sc. (1950) and D.Sc. (1958) from the University of Tokyo. He has taught at the University of Tokyo (1954-63) and the University of Chicago (1963-68) prior to joining the faculty at the University of California. Dr. Satake is a member of the Mathematical Society of Japan, the American Mathematical Society, and the Société Mathématique de France.

Design by Mary Ann Rosenfeld Liebert
Printed in the U.S.A.

ISBN: 0-8247-1607-8

MARCEL DEKKER, INC., NEW YORK